A Natural History of Chaos
An Introduction to Universal Darwinism

Michael Ent
Associate Professor of Psychology
Towson University
ment@towson.edu

A Natural History of Chaos
An Introduction to Universal Darwinism

Michael Ent

IBSN:
978-1-300-65882-5
Imprint: Lulu.com

Contents

Preface

In 1859, Charles Darwin proposed that new species evolve through a process of variation and selection.[1] Since then, scholars have filled in the details of this process and incorporated evolutionary principles into many disciplines. The idea that variation and selection explain phenomena beyond evolutionary biology is sometimes referred to as *Universal Darwinism*. This book provides a sketch of the myriad applications of Darwinian thinking.

I start with a discussion of basic evolutionary theory and how genes shape bodies and minds. Then, I discuss how Darwinian processes undergird trial-and-error learning, mental simulation, innovation, and cultural evolution. This book also explores the potential role of variation and selection in shaping the cosmos. Chaos is a topic of special focus as it is a fundamental source of variation.

This book is meant for a general audience. With that goal in mind, I include minimal technical language, which often means erring on the side of simplicity rather than precision. However, the notes and references provide suggestions for both technical and general further reading.

Darwinism helps to explain our world better than any other scientific framework. It is "capable of cutting right to the heart of everything in sight." [2] This book is an introduction to the role of Darwinian processes in shaping our world.

Chapter One
<u>The Emergence of Replicators</u>

With chaos and sufficient time, complex order can emerge. Four billion years ago, the world was without pain or pleasure – void of life. Chaotic interactions of chemicals in a "primeval soup" stumbled onto a combination that made copies of itself. This was the original replicator. Not all of the copies perfectly matched the original. Purely by chance, there were mutations. Some of the copies differed from the original replicator in their efficiency at creating more copies of themselves out of the primeval soup. Competition ensued. The efficient varieties of replicators proliferated; the others vanished. In other words, nature selected the fittest replicators. Over eons, variations of these replicators competed with each other for existence. This competition had cumulative effects. A single mutation made one variety of replicators less permeable, less vulnerable to the outside world. Those replicators with protective barriers would be especially adept at continuing their business of reproducing themselves without being destroyed. Replicators who had mutations that led to ever hardier bodies won against the others. Other mutations enabled these bodies to better navigate the environment, further aiding the ability of the replicators they housed to continue to reproduce. These bodies eventually became the fish of the sea, the birds of the air, the human race. The genes in every living thing are the descendants of the ancient replicators.

No one knows the exact origin of the first replicator, but the "primeval soup" account a reasonable guess.[1] The chemist Grahm Cairns-Smith argued that the original replicators spawned from clay crystals.[2] Replicators may have emerged in sulfuric thermal vents at the bottom of the ocean.[3] Proponents of panspermia believe that life came to earth on a meteor or other space traveler.[4] By this account, the original replicator would have been extraterrestrial, perhaps originating in the primeval soup of some other planet. Charles Darwin ignored the question of how life arose, other than speculating that took place in a "small, warm pond." One doesn't need to know the nature of the original replicator to understand the process that could unfold after one breaks onto the scene. If mutation occurs, a variety of replicators will emerge. If they are forced to compete for limited resources to continue their replication, then the most efficient variations will proliferate, and the others will die out.

The emergence of a replicator, as miraculous as it might seem, is insufficient for evolution to occur. What if the original replicator made perfect copies of itself every time? It could make a lot of the primeval soup into its own image, and the result would be akin to pond scum. The universe contains enough chaos to ensure that would not happen. Aberrations in the copying process are inevitable, and those aberrations make evolution possible. Chaos enabled both the emergence of the first replicator and the first mutation.

The Ingredients of Evolution

Evolution requires *replication, variation, and selection.* On his *Cosmos* television series, Carl Sagan used these principles to explain how Heike crabs came to have samurai warrior faces fashioned on their backs. The Heike were Japanese warriors who faced a brutal defeat during the naval battle of Dan-no-ura. Following this battle, fishermen became apprehensive about eating any crabs with shells that bore the slightest resemblance of a human face, a fallen warrior. When the fishermen would catch one of these Heike crabs, they would throw it back and it would be free to reproduce (to *replicate* its genes). The crabs with just an ordinary shell would be eaten, and therefore be prevented from reproducing crabs with ordinary shells. Because of genetic inheritance, the Heike crab's offspring would also be likely to have the samurai resemblance. Eventually, the sea would be full of Heike crabs mating with each other and making more offspring that look like them. Offspring are not perfect copies of their parents. Chance mutations and the shuffling that occurs due to genetic recombination in sexually reproducing species (*variation*) would cause some of the crabs to have more of the samurai resemblance than others. Those with the greatest samurai resemblance would be especially likely to be thrown back and have the opportunity to reproduce. Through this process, the backs of Heike crabs would increase in their resemblance of samurai warriors. In this case, humans acted as the *selection* mechanism, because they decided which crabs would survive and which ones would be eaten.

Humans act as a selection mechanism in animal breeding. By chance, some sheep are born with wool that is stronger and more lustrous than others (i.e., they display variation). If a sheepherder selectively breeds only those with the most desirable wool, subsequent generations of sheep would have increasingly lustrous wool. Through selective breeding, humans have transformed wolves into the vast array of modern dog breeds. Humans selectively breed dogs for traits ranging from hunting or herding ability to gentleness or cuteness. Darwin was inspired by the work of animal breeders when he formulated his theory of evolution through natural selection. He recognized that nature works as a selection mechanism.

Because of limited space and limited resources, not all organisms are able to survive and reproduce indefinitely. Nature may select for ability to find food, to evade predators, and the like. However, the ability to reproduce is the ultimate selective force in evolutionary biology. What would happen if no crabs would want to mate with a crab that had a samurai-looking shell? If such a crab arose by chance, it might be especially likely to avoid death at the hands of fishermen, but its genes would not be passed down if no crabs would mate with it. It would eventually die, and the sea would not be full of Heike crabs. In the evolutionary struggle, survival is only important insofar as it aids the reproduction of one's genes. Nature selects for the most efficient replicators.

Chapter Two
Blind Variation and Selective Retention

Evolutionary biology provides the clearest picture of a process found throughout nature. Biologists were just the first to spot it. Although the term *Universal Darwinism* has been used in a variety of ways,[1] it captures the idea that evolutionary principles extend beyond biology.

The most succinct description of Darwinism that doesn't explicitly evoke biology comes from the psychologist Donald Campbell. He called it *blind variation and selective retention* (BVSR).[2] BVSR encompasses the key ingredients of biological evolution. Chance mutations within genes are a form of *blind variation*, and those genetic variants that are most effective at replicating themselves are *selectively retained*. The relationship between BVSR in biology to BVSR in other disciplines is more than analogy.[3] It is different than comparing life to a winding path or a box of chocolates. Instead, BVSR is a abstract framework that explains wide-ranging phenomena. The purpose of this book is to outline the role of BVSR in myriad systems ranging from the human mind to the cosmos as a whole. First, it is necessary to outline some details about evolutionary biology.

Chapter Three
Genetic Evolution

The theory of evolution by natural selection is often summarized by the phrase "survival of the fittest." Which raises the question "the fittest what? the fittest organisms? the fittest species?" In educational materials, one might find the notion that animals behave in ways that are in the interest of the species. But natural selection happens within species. Organisms designed to be successful at competing with others in their own species would reproduce more efficiently than those who focused on the interest of their species. Richard Dawkins clarified in his book *The Selfish Gene* that nature selects for the fittest genes.[1] Dawkins urged taking a "genes'-eye" view of evolution. This view helps to explain why parents throughout the animal kingdom give up their lives for their children. If nature selected for the fittest organisms, this would make no sense. But children share a substantial portion of their genes with their parents. A mother sacrificing her life for her child would help ensure that the child would be able to pass on her genes to future generations. In general, people are more willing to sacrifice themselves for genetic relatives than non-relatives.[2] From the gene's-eye view, it matters little whether the genes are housed one body or the other. Genes use bodies to replicate themselves. Evolution by natural selection is about the survival of the fittest genes.

If you were to ask a mother why she would sacrifice herself to save her child, I doubt the answer would have anything to do with genetic inheritance. The answer would likely be love. Evolutionary psychologists distinguish u*ltimate* explanations from *proximate* explanations.[3] If I were asked to explain why I ate a sandwich, I might give the proximate explanation that I was hungry. The ultimate explanation would be that my body has minimum caloric requirements that I was satisfying. The proximate explanation for why people have sex is often that it feels good. The ultimate explanation would be that sex is necessary for reproduction and the passing on of one's genes. (Of course, sometimes people have sex explicitly for the purpose of procreation.) In the same manner, the proximate explanation for a mother sacrificing her life for her child could be love, while the ultimate explanation is the preservation of her genes. If the child has a greater chance of future reproduction than the mother, then this is a good tradeoff from the gene's-eye view.[4]

The concept of selfish genes doesn't imply that humans (or other organisms) are selfish. A mother sacrificing herself for her child is not selfish. Humans often help non-relatives as well. This can be adaptive from a gene's-eye view; people can aid their own survival and reproduction by forming cooperative relationships with others. These relationships are often characterized by reciprocal altruism (i.e., "If you scratch my back, I'll scratch yours.").[5] Blood-sharing among female vampire bats provides a useful example. Vampire bats who have just eaten will sometimes regurgitate blood into the mouths of hungry bats with whom they have established a cooperative relationship.[6] When one bat finds food but her friend doesn't, she may share with her friend. The friend may return the favor in the future. Humans form similar types of mutually beneficial relationships. While studying a group of modern hunter-gatherers, the anthropologist Lawrence Sugiyama found that a majority of adults had suffered an injury or illness that would have led to death if they had not been fed by others.[7] We all find ourselves in need of help at some point; having cooperative relationships with others helps ensure we will find help when we need it. Ultimately, such relationships aid survival and reproduction.[8] Selfish genes can give rise to unselfish behaviors.

Humans typically don't have the conscious goal of replicating their genes. If they did, then men would flock to sperm donation sites. That strategy would likely be more effective at spreading one's genes than starting family. There are rare cases of men adopting this type strategy, such as one fertility doctor named Paul B. Jones who secretly used his own sperm to artificially inseminate 12 women.[9] He was eventually caught and ordered to pay millions of dollars in a lawsuit. Evolution works too slowly to select specifically for sperm-donation behaviors, and our genes instill in us proximate goals (like sexual desire) that serve their ultimate goal of replicating themselves. Like other organisms, we have adaptations that aid our reproductive success (i.e., our fitness), even if we don't have the conscious goal of reproduction. In other words, "organisms are adaptation executers, not fitness pursuers." [10]

Obviously, genes don't "want" to replicate themselves in the way that humans might want things; they are the product of a blind process of variation and selection. Humans often have desires that differ from what is in the best interest of our genes. Reproducing as much as possible might be in our genes' interest, but not our own. Many people don't even want to have children. Our goals can transcend our genes' goal of replication.

Dawkins described our bodies as sophisticated vehicles for genes that serve the ultimate purpose of ensuring that the genes that are housed within survive and replicate.[11] The unique nucleotide sequences of your DNA that make up your genes are referred to as your genotype. They are recipes to build your organs, bones, muscles, neurons, and the rest of your body: your phenotype. Genes must work together to fashion a fit phenotype. For example, genes for large saber teeth wouldn't do a sloth much good, but they are quite useful for a fast and cunning tiger. Genes also contain recipes for creations outside the body. For example, there are genes in birds that direct them to build nests (i.e., nest building is innate for many birds). Nests help birds to survive and ultimately pass on their nest-making genes. Dawkins refers to such creations as the *extended phenotype*.[12] As I will cover later, the human genetic endowment includes adaptations that facilitate participation in cultural systems. For this reason, Steven Pinker and others have noted that culture is part of the human extended phenotype.[13]

Individual bodies will die and nests will disintegrate, but genes can be passed on indefinitely. For this reason,

Dawkins had considered titling his landmark book *The Immortal Gene* (rather than *The Selfish Gene*).

Genes Shape Who We Are

When geneticists say that genes influence a trait such as height, this might give the false impression that there is a single "height" gene encoded in one's DNA. In reality, most traits are influenced by a constellation of different genes working in concert.[14] Modern genetics research has made progress in cataloging the complex interactions of genes that give rise to physical and psychological traits. Nevertheless, researchers can estimate the degree to which a given trait is influenced by genes without identifying the specific genes responsible. This involves comparing the relative influence of genes vs. environment. Two such methods are adoption studies and twin studies.

In adoption studies, researchers compare how similar children are to their biological parents (the source of their genes) vs. their adoptive parents (the source of their home environment). If the adoptees are more similar to their biological parents than their adoptive parents on a given trait, then this suggests that genes have a larger effect on the trait than home environment.

In one type of twin study, researchers compare monozygotic (i.e., identical) twins to dizygotic twins (sometimes known as "fraternal" twins). Identical twins share 100% of their genes; fraternal twins share some genes, but only to the extent that any siblings share genes. In other words, identical twins are more genetically similar than fraternal twins. All twins who are raised together more or less share an environment (they even share a womb). If genes were inconsequential, then identical twins would be no more similar to each other than fraternal twins. If identical twins are more similar on a given trait than fraternal twins, then this suggests that genes play an important role in shaping that trait (i.e., the trait is heritable).

Researchers have also studied identical twins who were raised separately. One such study uncovered the striking case of the "Jim twins." In the 1940s, identical twin boys were adopted into different homes. By coincidence, both of their adoptive parents named them Jim. When they reunited more than three decades later, they noticed many uncanny similarities. They both married twice, first to women named Linda, then to women named Betty. Both previously had dogs named Toy. They smoked the same brand of cigarettes, drank the same brand of beer, and drove the same model of car. Both men had a carpentry hobby and nail-biting habit.[15] Anecdotes such as the Jim twins could be dismissed as bizarre coincidences. However, a large-scale study by Thomas Bouchard and colleagues compared identical twins who were raised together vs. raised apart on many dimensions, including personality, occupation, hobbies, and social attitudes. They found that identical twins raised apart were "about as similar" as those raised together.[16]

Many decades of research on behavioral genetics, employing various methods, suggest that genes play an important role in shaping who we are.[17] Erik Turkheimer summarized some important take-away messages of behavioral genetics in three laws: 1) "All human behavioral traits are heritable," 2) "The effect of being raised in the same family is smaller than the effect of genes," and 3) "A substantial portion of the variation in complex human behavioral traits is not accounted for by the effects of genes or families." [18]

Turkheimer's first two laws highlight the importance of genes in shaping people's personalities, talents, interests, and the like. The third law raises an interesting question: If people vary for reasons other than their genes or families, then what causes such variation? One answer is that there are environmental factors other than families. For example, identical twins raised in the same home might still have different teachers or different friends that influence them. Another relevant, and often ignored, factor is known as *stochastic developmental variation*.[19] (Stochastic means randomly determined.) There is an element of chance in how one's genotype will be expressed as one's phenotype. In other words, the process of turning genetic information into bones, muscle, and brains is chaotic. Hypothetically, two people could share the same genes and the *exact* same environment and still turn out differently. Traditionally, any variation in traits that can't be pinned on genetics is attributed to environmental influence. This approach ignores stochastic developmental variation and gives the impression that the environment is more influential than it actually is. This is one of many cases in which science sweeps chaos under the rug.

Gene-Environment Interaction

Separating genetic influence from environmental influence can be difficult because people's genes affect their environment. For example, as a conservative estimate, about half of individual differences in intelligence are due to genes (the other half is a mix of environmental influences and stochastic developmental variation).[21] According to the geneticist Robert Plomin "Children with a genetic propensity for high intelligence are likely to read books and select friends and hobbies that stimulate their cognitive development." He argues that we "select, modify, and create environments correlated with our genetic propensities." [22] People with a genetic propensity toward thrill seeking tend to occupy environments full of adventure. Those with a genetic propensity for extraversion may seek environments with many opportunities for social interaction, while introverts may avoid such environments. Our genes guide us to shape our environments.

Human Nature

The findings of behavioral genetics highlight the role genes play in shaping who we are. However, they tend to focus on how humans differ from one another. This book is more concerned with how our genes account for the ways humans are all similar. Genes shape human nature.

Chapter Four
<u>Ancient Behavioral Programs</u>

We are not born blank slates. Just as evolution fashioned bodies, it fashioned minds. Brains are a part of the "vehicles" genes use to perpetuate themselves. They come pre-equipped with some ancient programs. These programs guide migratory birds to fly south for the winter, salmon to swim upstream to find suitable spawning grounds, and baby sea turtles to crawl into the ocean after hatching. As Steven Pinker pointed out, if goats didn't have any inborn instincts, then they might attempt to hunt deer for food rather than grazing in a field; hens might eat their eggs rather than laying on them.[1] (Of course, instincts aren't perfect. One of my backyard hens is an occasional egg eater.) Humans also are born with reflexes and instincts. For example, the rooting reflex helps human babies to nurse. Some fears are also innately prepared. I tested this idea on my son when he was a baby by showing him a spider crawling on my arm. Despite no previous interactions with spiders, he got a terrified look on his face and began screaming. Actual empirical work has shown that human infants innately associate snakes with fear.[2] Similarly, research on chimps and monkeys shows that they have an innate propensity to fear snakes.[3] Some fears come from our genetic endowment. As noted by evolutionary psychologists John Tooby and Leda Cosmides, to have an innately prepared fear of snakes, one must have some conception of what a snake is before experiencing one.[4] The idea of such innate conceptions resembles a less mystical version of Carl Jung's idea of archetypes (i.e., mental representations shared by all people).[5]

In the same way that chance mutations can cause variations in physical characteristics, they can cause variation in behavioral tendencies. These tendencies are subject to BVSR. By chance, one of our ancient ancestors was born especially avoidant of snakes. Avoiding dangerous snakes could increase the chances that this ancestor would be able to pass on its snake-fearing genes to future generations.

One of our ancestor's friends may have been born especially avoidant of butterflies. This might be inconsequential or even detrimental to its success. Blind variation in behavioral tendencies leads to the selective retention of those that prove adaptive (i.e., they aid survival and ultimately reproduction). That you are alive today means that your ancestors had behavioral tendencies that were adaptive. You inherited these ancient psychological adaptations.

All humans tend to share a basic toolkit of psychological adaptations that we might call human nature. We tend to fear large predators, feel disgust at signs of contagious illness, and enjoy being accepted and loved by others (to name a few). But people differ in their propensity to feel fear, disgust, and need for acceptance. One reason for this is that our ancestors differed in what strategies worked for them. All humans have some instinctive fears, but some of our ancestors may have succeeded in the evolutionary struggle by displaying bravery, taking risks that paid off; others may have succeeded by playing it safe. In addition, behavioral tendencies that are adaptive in one environment may be maladaptive in other environments. There is no such thing as a perfect suite of human psychological adaptations.

Psychological adaptations are not always present at birth. The same goes for physical adaptations like teeth (having teeth is certainly adaptive, but human babies are not born with them). Human infants begin to display an innately prepared fear of heights, but this only shows up when they begin to crawl. And it shows up even if they have never fallen before.[6] Similarly, mating instincts kick into gear around puberty. Thus, some psychological adaptations are activated based on one's stage of development. They can also be conditional. For example, parental instinct may lay dormant if one doesn't have children, but they activate when one has a baby (with the help of hormonal changes).[7]

Evolutionary Mismatch

There is often a mismatch between the environments in which our psychological adaptations evolved and the ones we find ourselves in.[8] The first anatomically modern humans evolved in Africa around 200,000 years ago.[9] Behaviors and preferences that would have been adaptive in that environment can be troublesome in the modern day. For example, in Pleistocene Africa it would have been advantageous to gorge on any sugary food one came across, as this would be a rare opportunity to acquire much-needed calories. In modern environments, our evolved preferences for sweet foods can lead to unhealthy eating habits. There were no supermarkets full of ice cream and cheesecake during the Pleistocene. Ice cream is an example of a *supernormal stimulus* (a stimulus made by humans that activates our senses more than anything found in nature). Our evolved tendency to seek calorie-rich foods is hyperactivated when presented with ice cream, candy, and the like. A stark example of a supernormal stimulus involves the Australian jewel beetle. Female jewel beetles are brown with little dimples, features that the male beetles find sexy. Discarded beer bottles that wash up onto the beach often have these features in an exaggerated form (simmering brown with lots and lots of dimples). When male jewel beetles would find a beer bottle, they would ignore the female beetles and try to mate with the beer bottles.[10] This story has a happy ending in that companies changed the design of their beer bottles, and the male jewel beetles went back to mating with the females. There are limitations to the utility of ancient instincts and predilections. But they don't guide all behavior; some behaviors are learned.

Adaptations for Learning

The capacity to learn is itself the result of psychological adaptations. Learning cannot occur without specialized programs to guide the process. For example, in operant conditioning, an organism increases the probability of behaviors that are rewarded and decreases the probability of behaviors that are punished. There must be a behavioral program that says "if the behavior is rewarded, then increase that behavior." It is a simple "if X, then Y" command.

Operant conditioning also assumes that organisms are prewired to find some stimuli (like food or sex) rewarding. Further, research on operant conditioning shows that organisms are sensitive to *reinforcement schedules*. They modulate their responses depending on whether a behavior is rewarded every time it is executed vs. only every two times, or depending on whether it is rewarded once every hour vs. once every two hours. All of these contingencies would require behavioral programs specialized for learning.

In classical conditioning, organisms learn associations between stimuli that are presented together (like Pavlov's dog learning that a bell meant that food was coming). There must be a program that allows organisms to make such associations. Further, the Garcia-Koelling effect suggests that organisms are pre-programmed to learn certain types of associations.[11] Studies of conditioned taste aversion show that if you feed rats, then give them a dose of radiation to make them sick, they will learn the association between the food and the sickness. This will cause the rats to avoid that type of food in the future. However, rats don't make that same type of association if radiation sickness is paired with flashing lights and sound. Rats more easily learn to associate food with sickness than to learn to associate lights and sound with sickness. Evolved learning programs shape how we learn and the types of information we extract from the environment.

Evolved learning programs are shaped by BVSR. As with other psychological adaptations, they are affected by chance mutations (blind variation of genes), and nature selectively retains those with efficient learning programs. For example, in operant conditioning, organisms generally stop behaviors when they cease to be reinforced. This is known as *extinction*. You can train a rat to push a lever if you give it food after it does so. But if the rat presses the lever a bunch of times, and no food comes, it will tend to stop pressing. By chance, some ancestors of modern rats may have been more or less prone to extinction. Over time, nature selected for those with good balance of sticking with a behavior that paid off in the past vs. giving up on a behavior that is no longer paying off. Throughout history, organisms have had variations in their pre-wired learning programs. Nature selects those with adaptive learning programs

Resistance to Evolutionary Psychology

I have barely scratched the surface in describing how psychological adaptations guide behavior. Evolutionary psychologists have detailed how psychological adaptations regulate human aggression, social exchange, mating, and much more.[12] These phenomena manifest themselves in similar ways in other primates. Scientists seem to have no qualms with explaining chimpanzee mating behavior or aggression in evolutionary terms. However, many balk at turning the evolutionary lens on human behavior.[13]

Evolutionary psychology seems to have something for everyone to hate. Religious conservatives have resisted evolutionary thinking because it conflicts with the belief that God created humans. The 18th century theologian William Paley famously argued that if you found a watch on the ground, you would assume it was designed by a watchmaker.[14] How else could one explain the intricate arrangement of the gears and springs that underly the function of the watch? Paley argued that the appearance of design implies a designer, and that the biological world is full of evidence of design. How else could one explain the intricate structure and function of an organ such as the human eye? According to Paley, because organisms display evidence of design, they must have been created by God. Darwin's theory of evolution by natural selection provided an alternative explanation for the appearance of design in biology. As a result, it has been rejected by many religious people. As will be covered in Chapter 14, political progressives have discounted human nature because it places constraints on projects of social engineering. Many have reservations about evolutionary thinking because some of its early proponents attempted to justify human cruelty as a natural byproduct of the Darwinian struggle for existence.[15]

More generally, the findings of evolutionary psychology often clash with our intuitions about our relationships and ourselves. Evolutionary explanations may seem to steal the magic of romance or parental love. Nevertheless, evolution shaped both our bodies and minds.

Those who deny evolutionary explanations of human behavior face major obstacles. Take for example the finding mentioned earlier that humans in all cultures tend to help genetic relatives more than non-relatives. This is known as *kin altruism* and is found across the animal kingdom. It also follows straightforward evolutionary logic: organisms help those with whom they share genes because nature selects for the fittest genes. Our pre-human ancestors presumably displayed kin altruism. Why would nature wipe out that ancient behavioral programming in humans? If it did, how could one explain kin altruism popping up in human cultures across the world?[16] These questions are difficult to answer if one views all human behavior as the result of nurture rather than nature.

Human behavior is tricky to explain purely in evolutionary terms because we are influenced by culture more than any other animal. Understanding human behavior requires accounting for both nature and culture. The relationship between the two will be covered in Chapter 8.

Chapter Five
Blind Variation and Selective Retention (BVSR) Within Organisms

The psychologist Gary Cziko distinguished BVSR among organisms (genetic evolution) vs. BVSR within organisms.[1] As previously noted, the ultimate unit of biological evolution is the gene, rather than the organism. Nevertheless, this is a useful distinction. He used the Jerne-Burnet clonal-selection theory of antibody production as a fundamental example of BVSR occurring *within* an organism. Humans are born with an immune system that produces some antibodies to fight pathogens (i.e., disease-causing agents). However, they are not equipped with enough of a variety of antibodies to fight every pathogen they might encounter. How could they? Pathogens evolve alongside humans, and new one's pop up all the time (e.g., new variants of viruses).

The immune system's solution is to blindly create a variety of antibodies. When one of these randomly generated antibodies binds to pathogen and neutralizes it, the body produces clones of the one that proved successful. The immune system "remembers" those variants that bound to pathogens in the past so that they can be reproduced next time that pathogen is encountered. The immune system engages in blind variation of antibody production and selectively retains those variants that successfully fight pathogens.[2]

The immune system is a product of the BVSR of natural selection. A well-functioning immune system requires efficient use of BVSR *within* the organism. This is an example of the BVSR of genetic evolution selecting for organisms that effectively harness the power of BVSR within themselves.

As I will discuss later in this chapter, trial-and-error problem solving is also a process of BVSR that occurs within a given organism. Nature could not equip us with preprogrammed solutions to every problem we may encounter. When faced with a novel puzzle, animals blindly vary their behavior and selectively retain the behaviors that yield a solution. The first step is blind variation of behavior, which is the topic of the next section.

Unpredictable Behavior

The Harvard Law of Animal Behavior states that, even under "carefully controlled experimental circumstances, the animal behaves as it damned will pleases." [3] Although, this "law" may be a bit tongue in cheek, it alludes to the fact that some behaviors seem unpredictable in principle (not just unpredictable given our state of knowledge and observational techniques). In the next section, I will discuss the role of random behavior in allowing for adaptive use of BVSR within organisms. Blind variation of behavior can lead to the selective retention of those behaviors that lead to desirable outcomes. But random behaviors can be adaptive even without the selective-retention piece. Random behaviors can help organisms to evade predators and confound competitors.

Predictability makes one vulnerable to exploitation. The biologist and neuroscientist, Björn Brembs, described various examples of how seemingly efficient escape responses in animals can prove maladaptive if they are predictable.[4] Worms have an instinct to crawl out of the ground when they sense vibrations indicative of a mole in the area, because moles eat worms. Humans have taken advantage of this predictable behavior in the form of *worm grunting*. They make vibrations in the earth using metal rods, then wait for the zombie-like manifestation of worms emerging from the ground. Worm grunters then easily harvest the worms for fishing bait. Similarly, non-human predators evolve to exploit the predictability of their prey. Some snakes have evolved to manipulate the escape response of fish in a way that causes them to jump right into the snake's mouth. Brembs pointed out that flies evolved to have randomness built into their flight patterns. That is why it can be so frustrating to try and swat them. A touch of randomness in behavior can make for unpredictable escape responses.

Unpredictability can also be used strategically when humans compete with one another.[5] In many types of competitions, a by-the-book move may have objective merits but still be the wrong move if one's opponent can anticipate and counter it. In the words of chess champion Magnus Carlson, "having preferences means having weaknesses." [6] The fact that deciding on the best move often involves anticipating the responses of others is one of many insights from game theory, which was pioneered thinkers such as John von Neumann, and later, John Nash (whose life was dramatized the film *A Beautiful Mind*). From a game-theoretic perspective, unpredictability can be an effective strategy as it can confound competitors. For example, the 1999 Women's World Cup was won during a shoot-out by a right-footed player kicking the game-winning goal with her left foot.[7] Although she would have been able to kick the ball harder and more accurately with her dominant foot, the less predictable option of going with her non-dominant foot paid off. Unpredictability doesn't need to be strategically deployed to be beneficial. For example, the innately programmed random movement of flies aids their evasiveness. A dose of randomness in behavior can prevent predators and competitors from exploiting predictability.

Problem Solving and Learning by BVSR

Randomness is good for much more than making one unpredictable. Randomness can aid problem solving and learning. Clues as to how this could work come from W. Ross Ashby's book *Design for a Brain*.[8] In this pioneering book on cybernetics, Ashby attempted to figure out how the brain produces adaptive behavior. To illustrate some of his central ideas, he built a machine capable of regulating itself, which he called the *homeostat*. The homeostat starts at a state of equilibrium, then "stress" is introduced to the system to throw it off-kilter. What mechanism did Ashby devise to allow the homeostat to bring itself back into a state of equilibrium? He programmed it to randomly vary its settings until it found an equilibrium state, then to retain those settings (BVSR). The settings the machine tried out were based on Fisher and Yates' Table of Random Numbers. (As will be addressed in Chapter 15, this may constitute quasi-randomness, as "random" has a strict definition.) A dose of randomness was imposed on the homeostat in the form of "stress" to throw it into a state of disorder, and randomness used by the system in the form of BVSR was a means of restoring order. In more fanciful language, chaos imposed from the outside was overcome by harnessing chaos within.

Donald Campbell noted that this simple mechanism for self-regulation may seem like "something of a letdown." [9] However, Ashby demonstrated that random responses and the selective retention of those that yield desirable outcomes can be sufficient for problem solving.

Not all problem solving relies on BVSR. For example, multiplication problems can be solved by following a simple set of steps. Still, many problems we face cannot be solved in this fashion. BVSR is versatile enough to tackle problems when there are no tried-and-true methods for finding a solution.

Similar to Ashby's homeostat, organisms use BVSR to solve problems. Exploratory locomotion in protozoa is a primitive example of BVSR of behavior.[10] These single-celled organisms move forward until an obstruction is encountered, at which point they blindly vary the direction of their locomotion until they are able to resume forward movement. The direction of locomotion that allows forward movement is selectively retained. In this case, memory isn't necessary for BVSR to be useful. Even if the protozoa didn't retain any knowledge of the environment to guide future navigation, BVSR can still solve the problem of continuing forward motion despite obstacles. Similarly, Ashby's homeostat didn't learn from previous experience which settings were likely to achieve equilibrium. It started the process of BVSR from scratch each time it found itself in a state of disequilibrium.

Unlike Ashby's homeostat, many animals retain the results of previous experiences of trial and error to guide future behavior – they learn. E.L. Thorndike studied this phenomenon by putting cats in a *puzzle box*.[11] The box contained a trap door that could be opened by hitting a lever that hung from the top of the box. Thorndike placed a piece of fish outside of the trap door to further motivate the cat to escape. The cat's instincts would be little help in solving this novel puzzle. When Thorndike put the cat inside of the box, it would flail around in a seemingly random fashion to find a means of escape. At some point, purely by accident, the cat would hit the lever that opened the trap door. The cat would then dash out of the box and devour the fish. Thordike then put the cat back in the box over and over again, and he found that over dozens of trials, the cat gradually learned to escape faster. In these initial trials, the cat did not seem to have a "eureka" moment when it figured out the solution; it simply increased the likelihood of the behavior (hitting the lever) that aided its escape. Thorndike's cats blindly varied their behavior and gradually began to selectively retain the behavior that proved beneficial (i.e., that was reinforced).[12] In other words, they learned how to escape the puzzle box through trial and error.

Humans also learn through trial and error. Further, we can learn from the trial and error of others. Children can learn to avoid touching fire after being burned, but they can also learn to avoid fire after watching someone else get burned. Social learning enables people to benefit from the trial and error of others.[13]

Instincts are behavioral tendencies that result from the BVSR of genetic evolution. In contrast, learning occurs through the BVSR of behaviors within one's lifetime. In other words, learning involves BVSR *within* an organism. The behaviorist psychologist B.F. Skinner noted that the process of learning resembles natural selection: "Just as genetic characteristics which arise as mutations are selection or discarded by their consequences, so novel forms of behavior are selected or discarded through reinforcement." [14]

Neural Darwinism

Trial-and-error learning is a sort of behavioral Darwinism.[15] Blind variation of behaviors leads to the selective retention of those that prove beneficial. The biologist and Nobel laureate Gerald Edelman provided a framework called *neural Darwinism* that may help to explain how this type of learning stems from BVSR within the brain.[16] Edelman's writing is notoriously difficult,[17] but it is worth providing a broad sketch of how BVSR may work in brain development (even if the sketch misses the intricacies of Edelman's theory).

The brain stores and processes information in a web of connections among billions of nerve cells (neurons). Every human brain has a unique arrangement of these connections. The connections are not all prewired; they evolve through experience.[18]

Within the web of neurons, there are individual networks of intimately connected neurons that Edelman calls *neuronal groups*. These neuronal groups are subject to a process of selection. Importantly, multiple neuronal groups can serve the same function.[19] When a baby is learning how to grasp an object, multiple networks of neurons can generate the grasping motion. Some of those networks will prove successful and allow that baby to grasp the object; others will prove unsuccessful. The neural networks that prove successful will be strengthened and will be more likely to activate in the future; the unsuccessful ones will be weakened. Thus, the brain receives feedback from the environment after putting various neural networks to the test, and it rewires itself accordingly. This feedback is mediated by pain, pleasure, and other emotions.[20] For example, if the activation of a neural network leads to a behavior that results in pain, that neural network will be weakened and therefore less likely to activate in the future. Blind variation of neural networks leads to blind variation of behaviors. The neural networks that result in beneficial behaviors are selected for. Although this discussion has focused on neural networks that guide behavior, Edelman argued that neural Darwinism also explains cognitive operations (e.g., categorizing objects).[21]

Neuronal groups are formed based on the blind variation of connections among neurons. Many connections that are formed prove useless or maladaptive and are discarded.[22] According to Gary Cziko, "by forming a large variety and number of new connections, the brain can select the combinations that work best." [23]

The essence of neural Darwinism is that neural networks are created through the blind variation of connections in the brain, and those that prove useful in interactions with the environment are selectively retained. In Edelman's words, "variation and selection within neural populations play key roles in the development and function of the brain." [24]

Conclusion

Blind variation of behaviors can help organisms to throw off predators and competitors. More importantly, selective retention of blindly generated behaviors is the basis of trial-and-error learning. The theory of neural Darwinism sheds light on how the BVSR of neural networks may give rise to the BVSR of behaviors.

Chapter 6
Mental Simulation

Solving problems through trial and error can be dangerous, as serious errors could leave one injured or dead. However, an analogous process occurs within the safe proving grounds of the mind. Instead of engaging in BVSR of behaviors in the physical world, humans and some other animals can engage in BVSR using mental models of the world. Research on the *floating peanut task* suggests orangutans possess a rudimentary form of this capacity.[1] In one demonstration, a researcher placed a peanut in a plastic tube and observed as an orangutan unsuccessfully tried to retrieve the nut with her hands. After giving up on this strategy, she waited for several minutes before filling her mouth with water then spat it into the tube. The nut floated to the top, and she took her reward. Unlike Thorndike's cats, this orangutan seemed to have a "eureka" moment when she realized the solution. No one knows exactly what was going through the orangutan's mind during the *incubation* period before she thought of the solution, but the process likely resembles what humans do when faced with difficult problem. This process involves thinking "What if I tried this? What if I tried that?" until selecting a solution worth trying out. In other words, blind variation of alternative strategies in the mind can lead to the selection of one worth executing.

As an aside, it is worth contemplating just how "blind" this process really is. For example, it might not occur to the orangutan to try to wink at the peanut as a means of obtaining it. Similarly, animals might not have an equal likelihood of trying out every behavior in their repertoire if they found themselves in a puzzle box. The strategies might not be totally random. Animal behaviors may be guided by instincts that result from the BVSR of biological evolution or by previous experience with the BVSR of trial-and-error learning. However, if those strategies don't work, the animals may try increasingly random behaviors to find a solution.[2] As I will cover in Chapter 15, not even genetic mutations are entirely random. The important point here is that "blind" according to Donald Campbell's conceptualization of BVSR refers to the fact that one does not know which of the available strategies will yield a solution until trying them out and that there is an element of chance in which strategies will be tried.[3] Chapter 15 includes further discussion on the meanings of "blind," "chance," "chaos," and related terms.

A benefit of trial and error in the mind (vs. the physical world) is that one can often realize a given strategy is no good before trying it out. Errors can be confined to one's imagination. Nevertheless, a solution that seems correct in one's mind might not work in the real world. BVSR in the mind can help whittle down the number of solutions that one tries out in the real world. Thinking through alternative strategies in the mind, rather than executing them in the physical world, can save on energy, time, and in some cases, unnecessary risk.

Trial-and-error problem solving in the mind requires mental simulation. One would need to simulate the consequences of a given strategy to decide whether it is worth trying out. On a basic level, the peanut-seeking orangutan seems to be capable of running the simulation of "What would happen if I spit water into the tube?" On the other hand, humans are capable of creating rich internal worlds using mental simulation.

Mental Time Travel

Mental simulation can help people to generate potential solutions to problems that they face in the immediate situation. But much of human mental simulation involves imaginary time travel. About half of the time, people's thoughts deviate from the here and now.[4] We can simulate anticipated challenges, generate a variety of potential solutions, and select those that seemed to have the best outcomes in the simulation (or at least avoid those that would end in calamity). In this way, mental simulation is a form of BVSR. If people anticipate a tense conversation with their spouse or a business negotiation, they tend to mentally simulate the interaction beforehand. They may run through various scenarios of what they might say and how the other person would respond. This could guide their approach to the actual interaction. People can selectively retain the responses that worked well in the simulation.

Simulating social interactions requires creating a mental model of one's interaction partner, with knowledge of their motivations and personality. This is no easy feat. Roy Baumeister and E.J. Masicampo reasoned that the human knack for mental simulation evolved due to the challenges of navigating social relationships and cultural systems.[5]

The utility of mental simulations depends in part on how well they map onto reality. A course of action that led to favorable outcomes in a mental simulation may not end well in the real world if the simulated world is nothing like real life. The accuracy of our mental simulations depends in part on our previous experience. For example, it would be difficult to accurately simulate the responses of a person we barely know.

In addition to simulating the future, we can mentally time travel to the past. Replaying previous experiences in our mind's eye can help us to extract more meaning from those experiences. During the replay, we may catch important details that eluded us at the time. Further, mental simulation allows us to replay *counterfactual* (i.e., hypothetical) versions of those experiences. After making a mistake, people often experience regret in the form of "if I had only done X, then Y would never have happened." People may have to generate a variety of hypothetical responses before arriving at one in which they could have avoided their mistake. This process is often unpleasant, but it is a means of learning from our mistakes. In contrast, when people generate counterfactuals about how the past could have turned out worse, they often feel grateful.[6]

Mental simulation allows us to generate a variety of hypothetical versions of the past and the future that can help guide our behavior. It frees our minds from the here and now.

The Tradeoffs of Mindwandering

Engaging in mental simulation requires letting the mind wander from the present moment. There are costs associated with doing so.[7] Mind wandering distracts people from completing tasks at hand, especially when faced with the boring, attention-demanding tasks that often come with modernity. When faced with a spreadsheet, the ghosts of our past may slip in through the ether. Our innate propensity toward mental simulation can sometimes be a sort of evolutionary mismatch with modern demands on our attention. In addition, people report being less happy when their minds are wandering than when they are in the moment (even when they are engaged in activities that aren't especially pleasant).[8] Gurus of many strips extoll the virtues of living in the moment. Modern research supports the idea that people are happier when they attend to the here-and-now compared to when their minds wander.

Being alone with one's thoughts allows for distraction-free mental simulation, but people generally don't like to be alone with their thoughts. Timothy Wilson and colleagues found that many people would rather self-administer electric shocks than sit alone in a room without stimulation.[9] The researchers first administered an electric shock to all participants and selected only those who reported that they would pay money to avoid receiving another shock. Then they told these participants that they would be placed in a room for 15 minutes to "entertain themselves with their thoughts." The researchers put the device to administer shocks in the room with the participants during that time. Despite the participants saying they would pay money to avoid being shocked again, 67% of the men and 25% of the women gave themselves at least one shock during the 15 minutes. One participant shocked himself 190 times. Being alone with one's thoughts is so aversive that electric shocks can seem like a welcome distraction.

People may avoid being alone with their thoughts simply because it can be boring. People differ in their propensity for mental simulation.[10] Being alone with one's thoughts may be especially boring for those with low levels of mental simulation. For those high in mental simulation, being alone with their thoughts can be unpleasant as it often involves contemplating past and anticipated problems. People may seek a life of constant distractions to avoid the mental simulations that arise when they are alone with their thoughts. Doing so prevents people from reaping the benefits of this core human capacity.

Part of what makes the human mind special is that it is not constrained to the here and now.[11] Although many human brain structures closely resemble those of other primates, those associated with planning and problem solving are much more developed in humans than our nonhuman relatives. Neuroimaging research suggests that when our minds wander, the most uniquely human parts of the brain are especially active. The authors of one relevant study noted that "apparently, when the brain/mind think in a free and unencumbered fashion, it uses its most human and complex parts." [12] The human capacity for creativity and problem solving stems from mental simulation, and mental simulation requires mind wandering. Although mindwandering comes with costs, it is a burden we must bear for our powers of insight.

Mental Simulation and Unconscious Goal Pursuit

Mental simulation allows for planning, which can reduce intrusive thoughts about unfulfilled goals and future obligations. These thoughts may pop into our heads while we are trying to relax. For example, during a dinner with friends, one might think "I need to check on that mysterious charge that keeps showing up on my credit card bill." Although they are annoying, these thoughts are the unconscious mind's way of reminding us of our goals. Importantly, a type of plan known as an *implementation intention* can reduce these intrusive thoughts.[13] An implementation intention is a plan in the form of "when X happens, I will do Y." [14] For example, "when I next check

my email, I will check on the mystery credit card charges."
The key is to tie the goal you want to accomplish with
something *specific* that is sure to occur in the future. Having
this type of plan stops the unconscious from reminding you
of the goal until the appropriate time specified in your plan.
There will be no spontaneous thoughts about the mysterious
credit charge until you check your email.

Implementation intentions not only reduce intrusive
thoughts about unfulfilled goals, they help people to
accomplish their goals. When people mentally simulate
exactly when and how one will accomplish a goal, automatic
processes step in to remind one of the goal at the appropriate
time.

Simulations Shape Behavior

Many studies have shown that conscious thought plays little
role in guiding our immediate behavior.[15] For example,
Benjamin Libet has shown that brain activity predicts
people's motor movements before they report a conscious
decision to act.[16] There is still much debate about the
implications of Libet's studies, but it seems that unconscious
brain processes can initiate actions independent of our
conscious intentions. However, this does not imply that
conscious thought is just along for the ride, with no impact
on our behavior. Baumeister and Masicampo argued that
conscious thought guides behavior by shaping our automatic
responses and attitudes.[17] Implementation intentions are one
such example in that conscious intentions shape automatic
goal pursuit.

Through mental simulation, the conscious mind can
shape the unconscious mind. We construct hypothetical
versions of the past and future, and the outcomes of those
simulations affect our responses in the real world. For
example, people sometimes develop intense fear of objects
and situations that they have not actually encountered.[18] If
one mentally simulates being trapped on sinking ship, then
the person may become afraid of going on ships – even if
they have never set foot on one. Mental simulations

involving ships can affect one's automatic response to ships. This happens because memories of simulated events are processed in a similar manner as actual experiences. Imagining an event can even cause people to mistakenly believe that it actually occurred.[19] Mentally simulated experiences and actual experiences both shape us.

Neuroscientific studies have found that imagining actions activate the same brain regions as actual actions.[20] Because of this, athletes benefit from mental practice. [21] A boxer can benefit from mentally simulating a fight and planning attacks and maneuvers. This simulation can shape his responses in the actual fight. In other words, imagined responses shape actual responses. The conscious process of mental simulation can shape automatic responses and attitudes; it can shape the unconscious.

Construction of Mental Simulations

Mental simulations of hypothetical scenarios are constructed from the blind variation of elements of our previous experience.[22] They are generated in a similar way as dreams, and both are examples of the brain making input for itself.[23] According to the activation-synthesis theory of dreams, during REM sleep an area of the brain called the pons generates random impulses. The forebrain, which is responsible for planning and problem solving, then tries to piece together these random impulses into a coherent story.[24] The fact that people often refer to mental simulations as "daydreams" highlights the similarity between the two. This doesn't mean that conscious mental simulations are totally random; we are especially likely to simulate scenarios that are likely or seem important given our past experience.[25] Dreams are often more random than conscious mental simulation. In the same way that conscious mental simulations lead us to puzzle over how we would respond to anticipated problems or how we could have acted differently in the past, Antti Revonsuo argued that dreams are functional in that that they allow us to simulate threatening situations to learn how to cope.[26]

Blind variation manifests itself in mental simulation in two ways: 1) in generating the simulations, and 2) in generating our potential responses to the simulated scenario. We can select strategies that proved advantageous in simulations to guide our behavior in the real world.

Suffering in Simulations

The simulations our brains create in the form of dreams tend to focus more on negative scenarios than positive ones.[27] This makes sense if dreams are for teaching us how to cope with threatening situations. Similarly, conscious mental simulations allow people to plan for problems that may arise in the future. This is often unpleasant. Moreover, many of the problems we simulate never arise, making the mental anguish that results from contemplating them pointless. In the words of Seneca "We suffer more often in imagination than reality."

Simulating hypothetical versions of past events can be beneficial because doing so allows us to learn from our mistakes. However, people often continue to ruminate about past problems long after they are able to gain a new understanding from doing so.

We may not have much control over the content of our dreams, but we have some control over which conscious simulations we entertain. For example, I have spent nights simulating how I would protect my family in the event of all manner of exceedingly unlikely, apocalyptic scenarios. Sometimes (but not always), I have the wherewithal to recognize the outlandishness of my mental simulations and guide my mind wandering to more fruitful pastures. Simulating threats that we never end up facing is often less harmful than failing to plan for a threat that does arise. Nevertheless, people can go overboard in simulating threats that have a low likelihood of arising.

Another common type of maladaptive simulation involves catastrophizing. After experiencing or anticipating a negative outcome, people may run simulations of how a minor problem could end in catastrophe. A student who is worried about a poor grade on an exam might think "If I get a bad grade on this exam, then I won't be able to get into good college, then I won't be able to get a good job, then I might not have enough money to start a family, then I might end up dying alone." This process makes molehills into mountains. We might not be able to stop our minds from generating catastrophic simulations, but research on cognitive-behavioral therapy suggests that people can learn to identify and make corrections to these maladaptive thought patterns.[28] Doing so is an exercise of metacognition (i.e., thinking about what you are thinking). If people can recognize unhelpful simulations, they can learn to override them.

Interrogating the utility and plausibility of blindly generated mental simulations can help people to select for adaptive thought patterns, or at least avoid the worst of them.

Simulating an Ideal Future

Not all simulations are negative. People sometimes fantasize about blissful futures. Such fantasies may motivate us to work hard to make them reality. They may also make our actual lives seem dull in comparison. More importantly, they are often based on inaccurate views about human psychology.

People may be able to simulate a future in which all of one's desires are met, and assume that such a life would be constant joy — a sort of happily ever after. However, that is not how the human mind works. Pleasure comes from satisfying needs, not from the absence of needs. Hunger makes eating pleasurable; thirst makes drinking pleasurable. The absence of these desires would rob one of the joy of food and drink. Humans also need oxygen to breath, but that need is typically fulfilled, so people don't derive pleasure from breathing. Our bodies generally take oxygen for granted.[29] To feel the pleasure of satisfying desires, people must first feel the sting of unfulfilled desires. Contrary to our expectations, a future in which all of our desires were fulfilled would be bereft of joy. Happiness comes from overcoming struggles, not from the absence of struggle.[30]

In our simulations, positive events may permanently raise our level of happiness, but in reality they seldom do. Some people have a happier baseline than others, but only extreme changes in circumstance tend to alter people's baseline levels of happiness.[31] Also, negative events are more likely to lower one's baseline happiness than positive events are to raise it.[32] We may feel good temporarily after a positive event, but our moods quickly return to baseline.[33] This phenomenon is referred to as the *hedonic treadmill.*[34] This principle transcends hedonic desires such as those for food and sex. When people accomplish a goal, they often feel good for a short period of time before setting their sights on a loftier one.[35] People can find themselves on a treadmill of thinking "if I only accomplish this next goal, I will finally have made it." This belief may be motivating, but the feeling of having "made it" often never comes. In other words, people's simulations of how they will feel after reaching some new milestone are often inaccurate. The hedonic treadmill makes the happily-ever-after ending perpetually elusive.

On a positive note, gratitude can help fight the hedonic treadmill. Favorable changes in one's life can continue to inspire joy if people stop to reflect on them even after the initial thrill has dissipated. Several studies have found that making people count their blessings makes them happier.[36]

Simulating the Self

To simulate interactions with others, we must simulate a version of ourselves. To simulate apologizing to a friend or asking a coworker for help, we need to conjure a vision of ourselves making the apology or asking the favor. We must have a sense of who we are and how we relate to other people. Our simulated versions of ourselves allow us to replay hypothetical versions of past interactions and plan our future actions. To run even the simple simulation of "if my car won't start tomorrow, I will call a tow truck" requires us simulate a future "I."

Multiple theorists have proposed that human consciousness evolved because it enables complex mental simulations including simulations of the self.[37] In the words of Dawkins, "perhaps consciousness arises when the brain's simulation of the world becomes so complete that it must include a model of itself." [38]

There are likely non-human animals that have a form of consciousness that does not include a model of the self. This is often referred to as phenomenal awareness.[39] According to the philosopher Thomas Nagel, "fundamentally an organism has conscious mental states if and only if there is something that it is like to *be* that organism – something it is like *for* the organism." [40] If there is something that it is like to be a bat, then a bat at least has consciousness in the form of phenomenal awareness. If there is nothing that it is like to be a bacterium, then bacteria don't have consciousness at all. Humans have more than phenomenal awareness of the world; we have a keen sense of our place in it. The ability to form a model of the self is a key feature of human consciousness.

The Self and Narrative Creation

A sense of self enables people to organize their experiences, emotions, and desires in a way that they can be easily communicated. We narrate our lives, which helps us to make sense of our experiences and share them with others. For example, while driving home from a stressful day at work, you may run through the day's events and form a story of how the day transpired (your thoughts and reactions are likely to take center stage). This process of narrative creation would help you to be able to share your experience with family or friends when you get home. A sense of self is necessary for narrating our experience.

When constructing narratives about our experiences, we tend to cast ourselves as the hero. At minimum, we tend to paint ourselves in a favorable light. This is unsurprising given that people have a desire for positive self-regard.[41] There are also social benefits to being able to finesse our personal narratives in ways that highlight our virtues and excuse our mistakes. In many cases, we may not be aware that we are doing so. To convince others of the favorable stories we tell about ourselves, we often must convince ourselves. In other words, self-deception facilitates interpersonal deception.[42] If people believe their own lies, then they would be unlikely to give off any clues that they were lying. In a previous section, I mentioned that people sometimes mistake simulated events for real events. If people replay an episode in their mind in a way that bends the truth in their favor, they may begin to mistake the favorable replay for the actual events. Forming personal narratives helps us to relay our experiences to others, but we often bend these narratives in self-serving ways.

Personal narratives help people to make sense of their experiences and find meaning. They enable us to stitch together the past, present, and future in a coherent fashion. However, people may avoid forming narratives about traumatic experiences, because dwelling on such experiences causes distress. Failure to process traumatic events can cause chronic, low-level stress that takes a toll on people's wellbeing.[43] Through decades of research, the psychologist James Pennebaker has found that encouraging people to write narratives about personal trauma aids physical and mental health.[44] In many studies, Pennebaker and his colleagues assigned participants to write about traumatic episodes from their past for 15-30 minutes each day for a few days. Other participants were assigned to write about neutral topics, like how they spend their time. After following up with the participants (in some cases, months later), the researchers found that those who wrote about trauma had fewer physician visits, better immune-system function, and more positive moods than those who wrote about neutral topics.[45] Forming personal narratives about traumatic events is stressful, but it can pay off in the long run.

The sense of self allows for the construction of personal narratives. These narratives help people to make sense of their experiences and communicate them to others.

Mental Simulation and Problem Solving

Like trial-and-error learning, mental simulation is an example of BVSR within organisms. Instead of trying out solutions to problems in the real world, we can try them out in the safety of our minds. Mental simulation frees us from the here and now. Simulating hypothetical futures can help us to plan. Simulating hypothetical versions of the past can help us to extract extra meaning from our experiences. Simulating past and future problems is often unpleasant, but it helps us to navigate our world. Humans have a unique ability to harness the power of BVSR in the form of mental simulation.

Chapter Seven
Innovation Through BVSR

Creative breakthroughs can occur through BVSR in the physical world. This can be observed among artists. For example, a 2021 documentary about the Beatles shows Paul McCartney strumming on his guitar in a seemingly aimless fashion until he stumbles on the structure of a song that soon became the hit "Get Back." Legendary record producer, Rick Rubin, relies on BVSR in the studio. Instead of having a grand vision of how a song or a record will turn out, he encourages artists to "try everything." [1] Then, he relies on his sagacious powers of selection to guide the work to greatness. Similarly, Picasso's sketchbooks are full of "false starts and wild experiments." [2] By selecting and refining the results of these experiments, Picasso changed the art world. As discussed in the previous chapter, trial and error in the mind (i.e., mental simulation) can substitute for trial and error in the physical world. Much of the work of artists and scientists takes place in their minds; this mental work often relies on BVSR.

Combinatorial Play

In his book "Origins of Genius," Dean Simonton studied the accounts of famous artists and scientists describing their creative process and found consistent allusions to BVSR in their thought patterns.[3] One of the most poignant descriptions came from famed mathematician Poincaré: "Ideas rose in crowds; I felt them collide until pairs interlocked, so to speak, making a stable combination. By the next morning I had established the existence of a class of Fuchsian functions." [4] Einstein described a similar process as key to his productivity that he called *combinatorial play*.[5] BVSR in the mind can help to solve specific problems, but it doesn't need to be directed toward a specific problem to be useful. The blind exploration of the relationships between concepts in memory (i.e., combinatorial play) can yield new insights.

Breakthroughs only occur through combinatorial play if one has the requisite background knowledge. Geniuses tend to read widely, often in areas outside their own discipline.[6] Fruitful combinatorial play requires a rich network of concepts in memory and opportunity for those concepts to intermingle (i.e., an *incubation period*). The incubation period often must last a long time, because most of the ideas that result from the blind variation of combinatorial play are bad ones. In the words of Poincaré, "among the great numbers of combinations blindly formed by the subliminal self, almost all are without interest and without utility." [7] Chemist and Nobel Laureate, Linus Pauling said "you aren't going to have good ideas unless you have lots of ideas and some sort of principle of selection." [8] Even geniuses require many blind variations of ideas before finding one worthwhile (i.e., worthy of being selected). Having lots of ideas increases the likelihood of having some good ones.

Neuroimaging research suggests that when people are left alone with their thoughts, their brains buzz with activity.[9] Breakthroughs often occur when one is engaged in mundane activities like taking a bath, going for a walk, or preparing for a meal (tasks that don't require much mental effort). This goes for combinatorial play as well as solving specific problems. Actively trying and failing to solve a problem creates anxiety, which hinders creative thought.[10] During the tasks of daily life, people can mull over various combinations of ideas in memory. As I will discuss shortly, this process is often unconscious. Incubating one's thoughts while carrying on with life can also be useful for gaining new information to throw into what William James referred to as the "seething cauldron of ideas" [11] The most famous example of this is the tale of Archimedes who was stumped in trying to figure out if a local king's crown was pure gold. While slipping into his bathtub, he noticed the level of the water rise. He realized that the principle of displacement was the missing piece he needed to solve the problem. According to legend, Archimedes was so excited that he leaped out of his tub and ran naked through the streets of Syracuse yelling "Eureka!" ("I have found it"). Similarly, Guttenberg had the idea for reusable type to mass-produce Bibles, but he didn't figure out a means of printing until he contemplated the workings of a wine press.[12] That missing piece led him to invent the printing press.

It is difficult to know where inspiration will strike, but random inputs from the environment can furnish the missing pieces necessary for innovation. One must be on the lookout for inspiration. According to Louis Pasteur "chance favours only the prepared mind." [13]

Feeding the Cauldron of Ideas

Throwing random or unusual ingredients into one's seething cauldron of ideas can promote creativity. Robert Sobel and Albert Rothenberg showed artists pairs of images that varied in their themes and composition.[14] In one condition, the images were presented side by side. In the other condition, the images were superimposed to create one phantasmagorical image. For example, an image of nuns standing in a row was superimposed on an image of a horse race. After viewing the images, the artists were instructed to draw anything they wanted to. Experts rated the drawings of those who saw the superimposed images as more creative than those who saw the same images side by side.

Bizarre imagery in one's dreams can act as random inputs to drive innovation. For example, the inventor Elias Howe had a dream in which cannibals surrounded him waving peculiar spears, which had small holes in the points of the blades. The spears gave him the inspiration for a new kind of sewing needle, and led Howe to invent the lockstitch sewing machine.[15] Similarly, August Kekulé is said to have discovered the chemical structure of benzene after having a dream of a snake biting its own tail. (I have found that taking a bit of glycine before bed can boost the strangeness of dreams.)

Unconscious Processing

Many great thinkers, including Einstein and Poincare, describe consciously engaging in the blind variation of ideas, either to solve specific problems or to engage in combinatorial play. However, a similar process also occurs within the unconscious. Research on the transitive inference paradigm has shed light on how this works. Ellenbogen and colleagues briefly taught people a set of abstract rules to learn (A>B, B>C, C>D, etc.).[16] The researchers then tested some participants on their knowledge of these rules after 20 minutes, some after 12 waking hours, and some after 12 hours including a night's sleep. In other words, the researchers assigned different incubation periods. After their incubation periods, these three groups were equal in their ability to remember the specific rules they learned. However, the groups differed in their ability to make inferences based on these rules. For example, if you knew the rule "A is greater than B" and the rule that "B is greater than C", you could infer that "A is greater than C" – even if you were never given the explicit rule that "A" is greater than "C." The participants who had a 12-hour incubation period while awake were better able to make these types of inferences than those with a 20-minute incubation period. Those who had a 12-hour incubation period that included sleep did the best of all. Sleep did not make people any better at simply recalling the rules, but it made them significantly better at drawing new conclusions based on those rules. Even during sleep, the mind engages in combinatorial play – the blind exploration of how various bits of information in memory relate to one another. When stuck on a problem or deliberating a complicated decision, it helps to sleep on it.

According to the *capacity principle*, the unconscious mind is better at reorganizing large amounts of information than the conscious mind.[17] This is because the latter has severe constrains on how many bits of information it can process simultaneously. The conscious mind tends to focus on one thought at a time, but the unconscious mind can entertain many ideas at once.[18] In this case, "unconscious" doesn't only refer to sleep; while awake, our minds perform many operations without our awareness. Our conscious knowledge and attitudes can be affected by unconscious reshuffling of bits of information in memory. Thus, intuitions that arise from the unconscious are not snap judgments. They are often the product of extensive unconscious processing. Creative people often report that new ideas sometimes seem to pop into their minds fully formed. These epiphanies are also the products of a tirelessly churning unconscious.

Divergent Thinking

When striving for innovation, it helps to be smart. High intelligence is associated with being able to process many bits of information in the mind at once.[19] This aids conscious combinatorial play because one can play with lots of ideas at once. However, intelligence is insufficient for creativity; lots of highly intelligent people have low levels of creative achievement.[20] Creativity requires divergent thinking, which involves generating many alternative, novel solutions to a problem at hand.[21] In contrast, convergent thinking is characterized by a narrow search for one correct answer. People can be high in intelligence, yet low in divergent thinking.[22] One common test of divergent thinking is the *alternate uses test,* in which people are asked to think of as many uses as possible for a common object like a brick.[23] People high in divergent thinking generate more uses, and more original uses than those low in divergent thinking. For example, many people may list that you could use a brick as a doorstop, but only those high in divergent thinking are likely to list that you could warm it up next to the fireplace then place it in your bed to keep your toes warm on cold nights. As discussed in the previous section, having lots of ideas (resulting from blind variation) is often a precondition for having a few innovative ones in the mix.

Two factors that affect the number of ideas that people can generate through blind variation are: 1) the number of mental elements one is working with, and 2) the number of associations one can make among those mental elements. In this case, "mental elements" may refer to facts or procedures one has learned, autobiographical memories, or information about the environment that one is currently processing. Obviously, differences in memory capacity can affect the number of mental elements at one's disposal. But people also differ how selective they are about the information that they let into their minds. Much of the information that the environment affords is irrelevant to us, so people either consciously or unconsciously filter out much of it. Ignoring extraneous information can be helpful in that it can allow people to focus attention on pressing matters, but some useful information might be filtered out in the process. Highly creative people tend to notice important details that others miss. People also differ in their propensity to make a wide array of associations. Narrowing the scope of one's associative network can be helpful in that many potential associations are meaningless, and pondering them would be a waste of time and effort. For example, it would likely be pointless to ponder how the theory of relativity relates to making an egg-salad sandwich. But creative people often make important associations that others miss, like Guttenberg realizing how the mechanics of a wine press were important for mass-producing Bibles. An open attentional filter and a propensity to make lots of associations can aid the generation of many ideas through blind variation.

Positive Emotion

Positive emotion both increases people's scope of attention and propensity to make a wide array of mental associations. This is a key insight from Barbara Fredrickson's *broaden-and-build* theory of positive emotions.[24] Positive emotion acts as a signal that it is safe to explore. Research using eye-tracking equipment suggests that people broaden their visual attention when they experience positive emotion.[25] This includes frequently changing the focus of their gaze and attending to objects in the periphery of their visual field.

Positive emotion not only encourages people to take in more information from the environment, it helps them to process information from memory. According to Fredrickson, positive emotion leads people to "connect the dots between disparate ideas and thereby act creatively." [26] In other words, positive emotion facilitates combinatorial play.

As mentioned, mundane activities that don't require much effortful thought afford opportunities to incubate ideas and engage in combinatorial play. The broaden-and-build theory suggests that the best such opportunities are pleasant activities. Commuting counts as a mundane activity, but it wouldn't be an optimal incubation period if you were in a negative mood due to fighting traffic. On the other hand, if taking a walk or preparing a meal puts you in a good mood, then those activities may encourage creative insights.

The link between positive emotion and creativity is at odds with the popular image of the tortured artist. Some people may create great art in the throes of anguish, but it is not the norm. Even tortured artists are likely to be more creative during moments when their anguish subsides. If tortured souls were especially creative during bouts of intense emotional turmoil, then one might expect that suicide notes would be full of deep and inspired prose. Multiple studies analyzing the content of suicide notes have found the opposite. Suicide notes tend to be concrete and banal.[27] They are more likely to include specific instructions like "remember to give the cat her medication" than to opine about the nature of life and death. The author of one relevant study concluded that "suicide notes are *not* the insightful documents we would like them to be." [28] Negative emotion tends to lead to "tunnel vision," which inhibits creativity.[29]

The next two sections explore how innovation is, in fact, linked to "a touch of madness" and the experience of aversity. In both cases, negative emotion may be along for the ride, but it is not source of creativity.

A Touch of Madness

People with psychotic-spectrum disorders often tend to filter out less information from the environment and make a wider array of associations than others.[30] These tendencies facilitate the generation of novel ideas. Two common examples of these disorders are schizophrenia and bipolar disorder (though the tendencies in question tend to manifest themselves during the manic phases of bipolar disorder more so than the depressive phases). Those who ponder the nature of genius often confront the idea that "great wits are sure to madness."[31] History is replete with examples of geniuses who suffered from psychotic disorders.[32] However, such disorders are often sufficiently debilitating that they interfere with one's ability to create masterful works.

Being inundated with extraneous stimuli due to a leaky filter can be overwhelming. Similarly, making tenuous associations can feed delusion (e.g., conspiratorial thinking). Having a wide-open attentional filter and a propensity to make a wide array of associations can aid the blind variation of ideas, but creativity also requires a selection mechanism to sort the good ideas from the bad. Donald Campbell highlighted the necessity of a well-functioning selection mechanism when he said "The value of wide ranging variation in thought trials is of course vitiated if there is not the precise application of a selective criterion which weeds out the overwhelming bulk of inadequate trials."[33] Similarly, the French poet Paul Valéry noted that innovation requires two faculties "The one makes up combinations; the other choses, recognizes what he wishes and what is important to him in the mass of the things which the former has imparted to him."[34] Having tons of ideas can make the process of sifting through them difficult. In addition, psychotic-spectrum disorders are often characterized by delusional or illogical thinking,[35] which can interfere with the selection process. For example, delusions of grandeur might make every idea might seem like a good one. These disorders may also reduce the likelihood that one will apply selection in the first place. For example, Albert Rothenberg found that schizophrenic poets tend to avoid editing and selecting from their work.[36] Generating many ideas through blind variation doesn't do much good without rigorous selection. Psychoticism aids creativity, but some people inherit a high enough dose that it is debilitating.

People can inherit a dose of psychoticism without having a full-blown psychological disorder.[37] Both the personality trait of psychoticism as well as psychotic-spectrum disorders are heritable (i.e., they tend to run in families). Research has shown that having a close relative with one of these disorders is associated with creativity.[38] One such study involved comparing the Icelandic national mental-health records to a list of notable creative accomplishment among Icelanders. Relatives of people with schizophrenia were two to six times more likely than the rest of the population to find their names on a list of Iceland's creative elite.[39] This study is remarkable in that it included nationwide data, but other studies on the topic have also found links between creativity and a family pedigree that includes psychotic-spectrum disorders.[40]

With the right dose of psychoticism, people might be able to benefit from the ability to generate many original ideas, but still have a capacity to subject them to logical selection. Seneca said "There is no great genius without a touch of madness." Research suggests that a *touch* of madness seems to be the sweet spot for creativity.

Antifragility

In his study of the origins of genius, Simonton noted the connection between trauma and adversity early in life and achievement later in life.[41] It is well known that trauma can have lasting negative impacts such as the development of post-traumatic distress disorder (PTSD). However, research suggests that post-traumatic growth may be more common than PTSD. As many as 70% of people who experience trauma report some positive change in their life as a result.[42] In contrast, the rates of people who develop PTSD as a result of combat exposure, rape, natural disasters, and the like are

10-40%.[43] It is possible that some reports of post-traumatic growth are the result of wishful thinking. Believing that something good came from one's suffering could be comforting. Nevertheless, if we take the accounts of trauma survivors seriously, trauma often seems to lead to positive change. Post-traumatic growth is more than resilience (the ability to return to the status quo); it is marked by *improvement* resulting from adversity.

Nassim Nicholas Taleb's concept of antifragility helps to explain post-traumatic growth.[44] Antifragility is not merely robustness, or the ability to face stressors and remain unscathed; it characterizes systems that *benefit* from disorder. For example, the Greek hydra was a multi-headed, mythological monster; for every head that was cut off the hydra, it would grow two more. The hydra is an early example of the concept of antifragility in that attacking it made it grow stronger. The immune system is real-world example. As previously mentioned, when the immune system encounters a novel pathogen it creates antibodies based on blind variation, then clones the ones that successfully bind to and neutralize the pathogen. Then, it "remembers" those successful antibodies for future use. The more pathogens it defeats, the stronger it becomes. A lack of exposure to pathogens makes for a weak immune system. Exposure to new problems sets into motion the process of BVSR. The more problems one successfully solves, the more solutions one can selectively retain for future use.

Harm does not need to rise to the level of trauma to foster growth. Antifragile systems benefit from minor stressors. In biology, this is known as hormesis. For example, small amounts of poison can strengthen an organism and enable it to tolerate larger doses.[45] A similar phenomenon has been found with dietary restriction. Intermittent fasting causes the body mild stress, and the result is a variety of improvements in health and longevity.[46] A lack of mild stress can lead to deterioration. This happens with bones; they become brittle if they are not subjected to some strain. Astronauts who spend long periods in space experience significant bone loss due to being deprived of the constant strain of Earth's gravity.[47] Exposure to small stressors can make one better able to handles larger ones.

Even antifragile systems can be damaged by some forms of stress. Although pithy, Nietzsche's famous expression that "what does not kill me makes me stronger" is a bit of an overstatement. Catastrophic injury could leave one permanently disabled. Similarly, many people who experience trauma develop PTSD that overshadows the potential for post-traumatic growth. Many people are stifled by early childhood adversity, and we are less likely to hear their stories than those who went on the achieve greatness. Chronic small-scale stressors can also lead to weakness rather than strength if people don't have an opportunity to heal.[48] Antifragility requires adequate opportunities for a system to improve itself following exposure to stress.

In a previous section, I mentioned that positive emotion facilitates growth and innovation. This seems to clash with the notion of antifragility in that in that exposure to adversity causes negative emotion. Does positive emotion induce growth or does exposure to adversity? Both play roles. It takes time for exposure to adversity to cause growth. In the moment, adversity may trigger negative emotions such as anxiety and fear, which limit creativity and problem solving.[49] However, when the storm has passed, the experience of adversity creates the opportunity for growth. It allows us to separate winning strategies from losing ones, provides fodder for future mental simulations, and can strengthen our coping ability. Positive emotions allow us to explore and play with ideas; adversity puts those ideas to the test. Adversity forces the hand of selection.

Personal and professional growth doesn't come from experiences in which everything goes according to plan; growth stems from successfully navigating problems. Chaotic forces push antifragile systems to evolve. Human bodies and minds are antifragile.

Useful Ideas

An innovative idea must be original (otherwise someone would have already thought of it), but it also must be useful.[50] The ravings of a madman may be highly original, but they are unlikely to be useful. Scientific ideas are useful insofar as they help explain and predict real phenomena. Artistic ideas are useful insofar as they stir the heart or inspire the mind, and people differ widely in their tastes. Artistic ideas don't need to be proven true or false to be useful, they need to find an audience. Because of the difference in what makes for a useful scientific vs. artistic idea, Dean Simonton argued that originality matters more in art than science.[51] In either case, the utility of an idea is determined by being tested in the real world. A scientific idea that seems plausible in one's mind may prove incorrect when subjected to the unforgiving light of systematic investigation. An idea that an artist finds moving might fall flat when exposed to wider audiences.

The first process of selection occurs in the creator's mind, but to prove useful, an idea must be subjected to additional selection criteria. In reflecting on the myriad selection mechanisms an idea must face, Donald Campbell urged us to "Think of what a small proportion of thought becomes conscious, and of conscious thought what a small proportion gets uttered, what a still smaller fragment gets published, and what a small proportion of what is published is used by the next intellectual generation. There is a tremendous wastefulness, slowness, and rarity of achievement." [52]

In research on creativity, the usefulness of idea is determined by its reception by a consensus of minds.[53] The number of times a scientist's work is cited by others is a frequently used metric of eminence. Artists' eminence is often measured by how their work is received by critics and wider audiences.

In some cases, someone with an eye for selection can be enough to separate the wheat from the chaff for original thinkers who lack much of an internal selection mechanism.

Little kids often make countless "abstract" paintings and drawings. Although many would only be cherished by cherished their parents, by chance, some of the compositions can be stunning. (I have framed a few such artworks made by my kids when they were little, though I admit that I'm a tad biased). In many cases, the work of "outsider" artists is discovered and curated by people who recognize the merits of the work. For example, Henry Darger was an outsider artist who suffered from mental illness. Alone in his room, he created thousands of artworks and wrote a 15,000-page novel. He painted many haunting, panoramic battle scenes featuring children whose likenesses were inspired by advertisements he found while digging through trash. Although much of his work wouldn't have wide appeal, many of his paintings and collages are breathtaking and have been featured in exhibits around the world. The creation of original works through blind variation and the selection of those works through curation can occur in separate minds. People may have a talent for one of these skills but not the other. Artificial intelligence has become increasingly adept at creating artwork, and it has the potential for massive output. Before AI learns to hone in on the tastes of humans, curators can help to separate the wheat from the chaff. As I will discuss in Chapter 9, we all act as curators in cultural selection.

More works, more success

As previously mentioned, having lots of ideas is often necessary to have good ideas. Similarly, producing lots of creative works is often necessary to have some be successful. In his study of geniuses, Dean Simonton noted "research has consistently shown that the most powerful single predictor of reputation among both contemporaries and future generations is the person's sum total of contributions." [54] Even among the masters, most creative works fail to gain much traction. For example, a study of famous composers (including Bach, Mozart, and Beethoven) found that only 35% of their works continue to be actively played in today's classical music scene.[55] Because of changing sensibilities, inconsistent feedback, and difficulty in determining which qualities of a work determine its success or failure, creators are often unable to predict which of their works will be a hit. A solution is to create as many works as possible that satisfy the creator's own selection criteria, and then subject them to cultural selection (cultural selection will be covered in later chapters).

Conclusion

Blindly combining concepts in one's mind can yield novel insights. It helps to have lots of concepts in one's mind to play with. An incubation period is also necessary, so that the concepts have the opportunity to intermingle. Effortful thought isn't always necessary for innovation; incubation occurs when engaged in mundane tasks and even while sleeping. During the incubation period, random inputs from the environment can sometime provide the missing puzzle pieces that lead to innovation.

Most ideas that result from blind variation are useless. To have a few good ideas, it is necessary to have lots of ideas to sift through. However, having lots of ideas won't lead to innovation unless one has a capacity to separate the good ideas from the bad (i.e., a rigorous selection mechanism).

Chapter Eight
The Evolution of Culture

When explaining the behavior of non-human animals such as a squirrel or a mollusk, people find it perfectly reasonable to rely on principles of evolutionary biology. As briefly discussed in Chapters 3 and 4, the evolutionary lens can yield many important insights about human psychology as well. However, human behavior is driven by culture to such a degree that the lens of evolutionary biology often seems insufficient in its explanatory power. For centuries, scholars and laypeople alike have debated the prominence of nature vs. nurture in driving human behavior. Are we primarily the products of our culture or our genes? Both are certainly important. Further, they are intimately intertwined. As Roy Baumeister laid out in his book *The Cultural Animal,* humans evolved to participate in cultural systems.[1]

Humans don't have the speed of a cheetah, the eyesight of a hawk, or the resilience of a tardigrade. Our physical bodies are weak compared to many non-human animals. The reality television show *Man vs. Beast* from 2003 vividly demonstrated this. In one stunt, a 363-pound champion sumo wrestler was easily defeated in a tug-of-war competition by a 180-pound female orangutan. Despite physical weakness, humans have thrived in inhospitable environments all over the earth. A single human alone in the wilderness stands little chance of survival without the knowledge and cooperation afforded by culture. According to the biologist Theodosius Dobzhansky, "the key to man's evolutionary success is neither bodily strength, nor robust health, nor resistance to environmental insults. It is adaptability by cultural means." [2]

Culture is a key feature of the human evolutionary strategy in the same way as a nest is a key feature of a robin's strategy. Though human cultures may differ in many respects, the capacity to build cultural systems comes from our genes. In other words, culture is part of the human extended phenotype.[3]

There are examples of culture in non-human animals, and such examples can help to explain the essence of culture. Some dolphins have discovered how to catch fish using "mud nets." One dolphin acts as the "ring maker" and strikes the seafloor while swimming in a circle around a school of fish. This encircles the fish in a plume of mud. The fish try to escape by jumping to the surface only to find a group of dolphins waiting to snatch them out of the air.[4] Importantly, this is a learned strategy, not an instinct shared by all dolphins. One dolphin (or a group of dolphins) stumbled onto this technique and other dolphins picked it up through cultural transmission. Thus, one group (or pod) of dolphins might have this cultural knowledge, while another may not. Other animals also display evidence of culture, such as primates who share knowledge about finding and cleaning their food.[5] Culture is as system of information sharing.[6] Humans have harnessed the power of culture more than any other animal, and it is responsible for the heights of human achievement. Culture allows the innovations of one clever mind to spread throughout one's group, to other groups, and down through generations.

Of course, it takes a clever mind to generate innovations worth spreading. It also takes some level of intelligence just to pick up on the innovations of others. In other words, participating in culture requires intelligence. The two reinforce one other. This is sometimes known as the *cultural brain hypothesis.*[7] Once humans began to participate in cultural systems, those with smarter brains had a huge advantage.[8] They would be better able to reap the benefits of culture to solve problems related to survival and reproduction (which would help them to pass on their genes). In other words, culture created a selection pressure for intelligence. This led humans to become smarter over time and therefore able to create more sophisticated cultural systems. These more sophisticated cultural systems required greater levels of intelligence to navigate, and so human intelligence and the complexity of cultural systems increased in lockstep. Our big brains are for participating in culture.

Cultural innovation benefits both the innovators and those who copy them.

Innovators benefit from prestige and deference from others.[9] Those who copy them benefit from adopting useful skills and knowledge.

Cultural progress moves faster than increases in intelligence. There have been modest increases in human intelligence over time,[10] but such increases alone cannot account for dramatic escalation of human achievement over the last few millennia. Humans aren't that much smarter than they were in the days of Aristotle; there were plenty of geniuses among the ancients. However, our understanding of the world and our technology has dramatically improved, even over the last few hundred years, due to culture.

Human advances are driven by people working together in cultural systems. A person working alone wouldn't have been able to put a man on the moon. Large feats require the coordinated effort of many specialists sharing information. Sure, there are project managers, but they often need to rely on the expertise of those on their teams to supplement the knowledge that they lack.[11] Even when people seem single-handedly responsible for major innovations, they don't start from scratch. They build on the work of others. It is poetic to think of Darwin alone on a beach in the Galapagos Islands deriving the theory of natural selection merely by his powers of observation, but much of his theory was inspired by combining information he learned from experts in selective breeding with the theories of Thomas Malthus.[12] Human innovation is a team effort enabled by our innate proclivities toward culture.

Cumulative Culture

The real magic of human culture doesn't come from merely passing good ideas along, but from people building on the innovations of others. This is often known as *cumulative culture*.[13] The process relies on BVSR. After some ancient ancestor got the bright idea of sharpening a stick to use as a spear for hunting, others would copy the strategy and make spears of their own. However, some of the copies would inevitably vary from the original. For example, they might be longer or shorter than the original, or made from a different type of wood. Through many hunting trips, people would be able to observe which variations were more successful than others, and they would selectively copy the ones that worked best. The process could continue. Down the line, another clever ancestor may have had the idea to fasten a sharp stone to the end of the spear using dried animal tissue. Variations in the methods for fashioning the spearhead and attaching it to the shaft would allow for further improvements on the hunting tool. As summarized by Tennie and colleagues, "Human cultural transmission is thus characterized by the so-called 'ratchet effect', in which modifications and improvements stay in the population fairly readily (with relatively little loss or backward slippage) until further changes ratchet things up again." [14] Cumulative culture allows people to build on the innovations of others and is the chief driver of human progress.

These ancestors wouldn't need to know the details about why one spear worked better than the others (e.g., the aerodynamics); they just needed to know which ones worked best and copy and refine those designs. A similar dynamic seems to have been at play with the development of f-holes in violins. The sound holes in early violins were typically circle shaped. However, over centuries, craftsmen gradually altered the shape of sound holes until they settled on the f-shaped holes characteristic of modern violins, which sound much better than earlier designs. It is not at all obvious that an f-shaped hole would conduct sound better than a circle or some other shape. There is no evidence that a single craftsman introduced the f-hole as the result of an ingenious insight about acoustic radiation. Instead, craftsmen seem to have copied the designs of the best sounding violins. Errors in the copying process led to deviations that yielded even better sound quality, and those were then copied. The authors of an innovative study on this topic noted "By evolution-rate analysis, these changes are found to be consistent with mutations arising within the range of accidental replication fluctuations from craftsmanship limitations with subsequent selection favouring instruments with higher air-resonance power." [15]

BVSR in the real world is sufficient for some cultural innovation, but we also engage in mental simulation. Instead of haphazardly altering the design of a spear, an ancient human may have run the mental simulation of "what would happen if I used a sharp stone as the tip of the spear?" As noted in Chapter 6, BVSR in the mind can guide BVSR in the real world.

Modern manifestations of culture ranging from buildings to political organizations to cuisine are the products of cumulative cultural innovation. In many cases, the chain of innovation stretches back to the ancient past.

BVSR Drives Human Success

Before discussing some of the characteristics of cultural evolution, it is worth reviewing the ways organisms have evolved to use BVSR to increase their fitness. The immune system, trial-and-error learning, mental simulation, and culture all rely on BVSR. Organisms that use these forms of BVSR have advantages over others. Human success results from harnessing the power of BVSR more than any other animal. Culture is an especially potent example of this.

Chapter Nine
Memes

Cultural systems evolve in a process similar to natural selection. Blind variations of ideas and practices are subjected to cultural selection. In biological evolution, the unit of selection is the gene. Richard Dawkins proposed using the term *meme* (derived from the Greek root for "imitation") to refer to the unit of cultural selection.[1] Any idea or practice that spreads through culture is a meme. Examples include melodies, jokes, stories, scientific theories, methods of building boats, and recipes for cooking stew. Memes replicate themselves by spreading from one mind to another. We act as the selection mechanisms for memes by determining which ones spread. Culture created a new form of replicator – the meme.

In human genetic evolution, genes are transmitted exclusively from parent to offspring. They are passed down from one generation to the next, which is often referred to as vertical transmission.[2] Memes also transmit vertically when older generations pass down knowledge, customs, and traditional practices. In addition, memes can be transmitted horizontally (i.e., between people in the same generation).[3] This occurs when one shares and idea with friends, family, coworkers, or strangers on the internet. In cultural evolution, memes are transmitted from one mind to others. With an increasingly connected web of information sharing, memes can quickly travel from one mind to minds across the globe. Thus, culture evolution moves much faster than genetic evolution.[4]

Meme Selection

There are a variety of factors that may affect the degree to which a meme spreads (i.e., a meme's fitness). Memes may spread by aiding human survival and reproduction. If participation in culture is part of the human evolutionary strategy, then we likely evolved to select memes that aid our own fitness. As Daniel Dennett put it "we would not survive unless we had a better-than-chance habit of choosing the memes that help us." [5] One reason intelligence is necessary to reap the benefits of culture is that it helps people to select advantageous memes. Culture would not confer any fitness advantages if people didn't have some capacity to select memes based on how well they aided survival and reproduction. Thus, memes might spread if they helped people to find food, make shelter, raise children, and the like. One method of finding fitness-enhancing memes is to find successful others and pick up on their strategies. [6]

Memes could also spread because people mistakenly believe that they help them reach their goals. That is, humans are not fully rational in determining what memes help us reach our goals. For example, the meme of human sacrifice could spread if people believed that the practice would help to ensure a plentiful harvest. In addition, some genuinely helpful memes may not catch on if people fail to see their utility. The anthropologist Manvir Singh pointed out that sanitation practices like handwashing are often slow to catch on through cultural transmission. One reason is that the benefits of such practice are difficult to directly observe, and there is a time lag between the practice and the benefit. [8]

Manvir Singh argued that memes may spread if we judge them to satisfy our "subjective" goals, even if they don't increase our fitness. [7] Smoking a cigar, looking at the sunset, or sitting in a comfortable chair may fulfill human desires irrespective of their fitness effects. Thus, a meme for how to build a comfortable chair may spread because if fulfills our subjective goals.

Some memes thrive in spite of people judging them to be annoying, ugly, or even harmful. [9] For example, someone might get a bit of a song stuck in their head and inadvertently pass it along to others while humming it at work. These "earworms" are often annoying, but they are successful in replicating themselves just because they are catchy. [10] On a darker note, suicide can also spread via memetic transmission. Publicized suicides often lead to "copycat suicides." The suicide meme is especially likely to spread after celebrities take their own lives. [11]

One could take a "meme's-eye" view of cultural
evolution similar to the "gene's-eye" view of biological
evolution discussed in Chapter 3.[12] Memes will spread as
long as they are good at replicating themselves. Memes
don't need to be truthful or helpful to be successful
replicators; they may proliferate because they are attention
grabbing, memorable, or easy to relay to others. Memes may
exploit our fears and desires to further their replication.

The criteria for successfully replicating memes interact.
For example, if a helpful idea is also catchy, it might be
more successful than an equally helpful idea that isn't. A
rhyming maxim like "haste makes waste" is more likely to
be transmitted than the less-snappy phrase "it helps to take
your time" (even though the message is the same).[13]

People may select memes that help them (or seem to
help them) reach their goals. But memes are selfish in the
way that genes are. They replicate themselves by whatever
means prove successful, and they use us to do so. As with
genes, memes don't "want" to replicate themselves the way
we may want to eat a piece of cake. They are simply
replicators subject to blind variation and selective retention.

Meme Curation

Some memes go viral in a bottom-up fashion, spreading
through social networks by literal or digital "word of
mouth." There are also meme curators who select songs,
artworks, scientific ideas, and the like for wide
dissemination. Meme curation can be helpful, because it
wouldn't be feasible for people to sift through every possible
meme to find the useful or entertaining ones.[14] For example,
people may seek the reviews of a trusted movie critic to
decide what to watch on a Friday night. However, meme
curation often occurs whether people want it or not, and
meme curators often don't act in the public interest.

Some people have more influence than others in
determining which memes will spread. People with wide
audiences, such as celebrities or political leaders, are
especially powerful in disseminating memes.[15] Regardless of

whether a meme is truthful or helpful, it can proliferate merely by being communicated by an influential person. Powerful people may be motivated to promote memes that make themselves look good or make their opponents look bad, even if those memes are false or dangerous.[16]

Social media platforms may also be motivated to spread harmful memes. These platforms promote content that drives engagement (i.e., keeps people on the platform). For social media companies that rely on advertisements for revenue, more engagement means more money. There are multiple ways to drive engagement. People may stay on a social media app to watch cute dog videos, chat with friends, or to voice their outrage at the latest political news. However, content that evokes negative emotions such as outrage is especially engaging.[17] This is because people attend to and are impacted by bad information more than good information. In other words, "bad is stronger than good." [18] There is evolutionary logic behind people's tendency to pay special attention to potential threats. In the words of John Tierney and Roy Baumeister, "To survive, life has to win every day. Death has to win just once… the hunter-gatherers who survived were the ones who paid more attention to shunning poisonous berries than to savoring delicious ones. They were more alert to predatory lions than to tasty gazelles." [19]

Social media platforms selectively promote content that drives engagement. Because bad is stronger than good, this leads to wide dissemination of content that stokes fear, outrage, and other negative emotions. All emotions serve functions. Fear can help people to avoid danger, and outrage can enforce moral norms. Nevertheless, these emotions can be harmful in excess, especially when they are elicited by sensationalized stories. Memes are selfish, and harmful ones can spread on their own. However, meme curators can facilitate the spread of harmful memes if it benefits them.

Memes from Fiction

In action movies, a silencer can turn the sound of a gunshot into a whisper, and the pin of a grenade can be pulled out with one's teeth. In reality, that is not the case. Fiction may sensationalize psychological disorders and give false impressions of how medical and judicial systems work. We often realize that fiction doesn't map onto reality, but memes from fiction can have an insidious effect.

Memes acquired from fictional sources are often incorporated into people's beliefs about the real world.[20] The *sleeper effect* suggests that this happens over time.[21] As time passes, we may remember information from fiction, but forget that it came from a fictional source. One might think "I remember hearing somewhere that brushing your teeth every day is actually bad for your gums," but forget that this idea came from some nincompoop in a comedy. In this way, memes from fiction can lead to false beliefs.

Fiction can shape our mental simulations, especially when simulating unfamiliar situations. For example, if you are scheduled to testify in court, but have no experience in a courtroom, your simulations may have little to go on but your previous viewings of television courtroom dramas. These feeble simulations may shape our expectations and potentially our future behavior.

False or harmful memes may spread through fiction because they are entertaining. Over time, they can shape our view of the real world.

Conclusion

Humans host two forms of replicators: genes and memes. Both are forms of information storage and transfer.[22] Memes are the units of cultural selection. Like genes, they can be passed down through generations. Unlike genes, they can spread quickly throughout the world in a single generation. Humans select which memes spread. We may select memes because they increase our fitness or help us fulfill our subjective goals. Harmful memes may also spread by being efficient at exploiting our psychological weaknesses for their own replication.

Chapter Ten
BVSR within Cultural Systems

There is a hierarchy (or perhaps a genealogy) of various forms of BVSR. Genetic evolution gave rise to culture. Cultural evolution, in turn, gave rise to systems that harness the power of BVSR. Two such examples are science and market economies.

Science

Human culture has developed multiple methods for gaining knowledge about the world. In other words, there are multiple knowledge-acquisition memes. When the science meme began to spread during the Enlightenment, it often replaced other memes related to knowledge acquisition such as divination (i.e., seeking knowledge through supernatural means). Part of the reason for the spread of the science meme is that it yields predictions that are more reliably correct than other knowledge-acquisition memes.

Science relies on BVSR. According to Karl Popper "The growth of our knowledge is the result of a process closely resembling what Darwin called 'natural selection'; that is, *the natural selection of hypotheses*: our knowledge consists, at every moment, of those hypotheses which have shown their (comparative) fitness by surviving so far in their struggle for existence; a competitive struggle which eliminates those hypotheses which are unfit." [1] Popper famously outlined the importance of falsification in science. Scientific ideas must be capable of being proven wrong. A process of elimination based on systematic investigation can separate the fit ideas from the unfit. This is similar to unfit genes being eliminated from the gene pool. Scientists generate and test a variety of hypotheses, and those that are supported by evidence are selectively retained; unsupported hypotheses are jettisoned. BVSR of ideas in the scientist's mind may guide hypothesis generation. Hypotheses may be based on logic and previous research, but scientists don't know which ones will be supported until they subject them to investigation.

For science to progress, scientists must be effective at separating the fit hypotheses from the unfit. In other words, science needs a good selection mechanism. Scientists are responsible for judging the fitness of hypotheses, and they can be biased. For example, scientists may champion unfit hypotheses because they align with their political preferences. This is especially troublesome when the majority of a discipline shares a political ideology. In such circumstances, there may be few voices to condemn unfit yet politically popular ideas.[2] Although imperfect, science increases human knowledge through the BVSR of hypotheses.

Market Economies

Human culture has developed multiple economic systems. For example, in command economies, governments control production and set the prices of goods and services. On the other hand, in market economies, production and prices are based on supply and demand. Sellers (ranging from individual entrepreneurs to large corporations) compete with each other for the business of consumers. Although command economies still exist, market economies have spread around the world due to the prosperity they bring with them.[3]

Market economies rely on BVSR. Businesses offer a range of products and services, and consumers act as the selection mechanism. It is difficult to tell which products will sell. Markets are notoriously unpredictable (hence all the betting that takes place on Wall Street). Products and that don't sell will cease to be produced; business that aren't profitable will die out. Variation in products and businesses leads to the selective retention of those that meet people's needs and desires. As the economist Milton Friedman pointed out, a process akin to natural selection undergirds the evolution of market economies.[4] The picture is complicated in that governments play a role in producing the conditions for market economies to thrive[5] (e.g., laws against monopolies). Large-scale economic systems are too complex for any top-down governing body to run successfully, though governments can set the stage. Market economies produce more prosperity than the alternatives that have been tried, and they are based on BVSR.

The Efficiency of BVSR in Cultural Systems

Human culture has given rise to multiple economic systems and multiple means of knowledge acquisition. Market economies and science have proven more successful at spreading than their rivals, and they are both are based on BVSR. In general, systems based on BVSR tend to be more efficient than alternatives.

Chapter Eleven
Cultural Differences and Similarities

All humans face a set of common obstacles to overcome. These include finding food and shelter, avoiding predators and pathogens, and the like. Some cultural differences arise because people have stumbled onto different solutions to these common problems. For example, all people need potable water, and there are multiple methods of preparing beverages that remove potentially harmful pathogens. Brewing beer and tea both involve heating water, which kills germs. People may discover that drinking beer or tea instead of untreated water lowers the risk of illness. (They wouldn't need to know that the boiling process was key.) Either solution would work, but some cultures have historically been tea drinkers and others beer drinkers.[1]

In many cases, cultures stumble onto the same solutions to common human problems. This is sometimes referred to as *convergent cultural evolution.*[2] In biology, convergent evolution occurs when different species independently evolve similar features. For example, human eyes and octopus eyes share many similarities, but octopuses and humans didn't get their eyes from a common ancestor; they evolved independently.[3] There are only so many efficient ways for animals to detect light, and nature stumbled onto a similar solution multiple times. The fact that many cultures around the world use bowls is likely the result of convergent cultural evolution rather than cultural transmission (or some "bowl-making" genes common to all humans). There are only so many ways to hold liquid, so it is no surprise that cultures around the globe independently discovered bowls as a solution. In different cultures, bowls may be made of different materials or have different ornaments, but the general design tends to be similar because they serve the same function. The bowl-making meme would not need to spread around the world via cultural transmission for bowls to be a common cultural artifact. Similarly, pyramids are a simple way to make a tall structure and they are found all around the world. This can be explained by convergent cultural evolution rather than cultural transmission.

Cultural Differences from Ecological Differences

Ecological conditions such as the physical terrain, temperature, or access to water can cause cultural differences.[4] Cultural norms and practices often evolve to suit specific physical environments, and humans inhabit many different climates and landscapes.

The types of crops that people can grow varies across the globe (e.g., it would be difficult to grow oranges in Siberia). Cultural differences arise from relying on different types of crops. Talhelm and colleagues argued that cultural differences between the north and south of China are driven by reliance on wheat farming vs. rice farming.[5] Rice farming requires significant cooperation between neighbors. Rice paddies require irrigation networks that neighboring families work together to maintain. The farmers must also coordinate their water use to ensure that each family gets their fair share. In contrast, wheat farmers can rely on rainfall rather than irrigation. Rice farming requires about twice the amount of labor as wheat. While wheat fields can be maintained on the labor of a single family, rice farmers must participate in labor exchanges to ensure that time-sensitive tasks like transplanting and harvesting are completed within a narrow window of time. Talhelm and colleagues found that the people in the rice-growing regions south of the Yangtze River are more collectivistic, display more holistic thinking, and value loyalty more than the people in the wheat-growing regions north of the Yangtze.

These cultural differences manifest themselves in subtle ways in everyday life. Observational studies of people in cafes found that those in the wheat-growing north were more likely to sit alone than those in the rice-growing south.[6] The researchers also went to Starbucks in five different Chinese cities and pushed chairs together to block an isle so that they could observe how people responded. They found that people in the north tended to move the chairs out of the way to get through, while people in the south tended to squeeze through the chairs. This is consistent with other work demonstrating that individualism is associated with adjusting the environment to fit the self, while collectivism is associated with adjusting the self to the environment. The researchers argued that these cultural differences emerged from the different interpersonal requirements for rice vs. wheat farming. Importantly, the researchers noted that the people they studied were not farmers. They noted "you do not need to farm rice yourself to inherit rice culture." [7] That is, cultural norms and practices that develop to deal with ecological challenges are passed down through cultural transmission.

Conclusion

Cultural similarities can arise by multiple cultures stumbling onto similar solutions to common human challenges. On the other hand, cultures sometimes find different solutions to these common human challenges, which leads to cultural differences. The challenges that humans face differ based on the environments they inhabit. Cultures evolve to suit their unique environments. Cultural differences can arise from environmental differences.

Chapter Twelve
Mismatches in Cultural Evolution

Chapter 4 described the phenomenon of evolutionary mismatch, in which some of our evolved tendencies can cause problems in modern environments. For example, cravings for calorie-rich foods that were adaptive in the ancient past may be harmful in modern environments where junk food is plentiful. There is an analogous phenomenon in cultural evolution.[1] Cultural norms and practices that evolved in one context may be harmful in other contexts.

In *honor cultures*, people use violence (or the threat of violence) to maintain a reputation of toughness. These cultures are likely to develop when there is a high risk of theft and low reliance on authorities to enforce order. Researchers have used the concept of honor cultures to explain the fact that, in the United States, the south tends to be more violent than the north. Specifically, southerners are more likely than northerners to respond to insults with violence. These differences have been documented in crime statistics, laboratory experiments, and surveys; and they persist when accounting for other variables such as demographics and temperature.[2]

In early settlements, herding was a chief form of agriculture in the American south, whereas farming was more common in the north. Herding incurs a higher risk of theft than farming. It would be conceivable for a man's livelihood to be stripped away at once if his herd were stolen. In contrast, thieves would be unlikely to harvest all of a farmer's wheat one night while he slept. In addition, until well into the 19th century, the southern US was a frontier region, and the government had little power to enforce the law. Thus, the early American south met the two criteria for developing an honor culture: 1) high risk of theft, 2) low reliance on authority to enforce order. In such circumstances, the threat of retaliation is the primary theft deterrent. Maintaining a ruthless reputation could help to dissuade potential thieves. On the other hand, being seen as weak or passive in the face of insults could mark one as a suitable target for plunder. In honor cultures, maintaining a ruthless reputation is a means of self-protection, and responding violently to insults helps to maintain such a reputation.[3]

The psychologist Richard Nisbett argued that the high rates of violence in the modern American south (vs. the north) are a holdover from a previous era in which southerners relied on the cultivation of a ruthless reputation as a means of self-protection.[4] The times have changed. The South is less reliant on herding, and the government has much more power to enforce laws. Nevertheless, remnants of the culture of honor persist due to cultural transmission.

As mentioned in the previous section, rice culture is passed down through cultural transmission to younger generations who don't farm rice. Similarly, cultures of honor may persist due to cultural transmission even when the conditions that led to them subside. The difference is that the collectivism resulting from rice farming is innocuous compared to the violence that results in transmitting honor culture. The propensity toward violence remains though it no longer serves the same self-protective function. For this reason, the transmission of honor culture from the early southern US to the modern day could be considered an example of a mismatch in cultural evolution. Cultural practices and norms that were useful in the past may be ill suited to modern environments.[5]

Evoked Culture vs. Convergent Cultural Evolution

American southerners are not the only ones who display honor culture.[6] Honor cultures seem to have independently sprung up in multiple areas around the world. There are two possibilities for how this could happen. One is convergent cultural evolution. That is, people might stumble onto the same strategy for self-protection when faced with similar conditions. The other possibility is known as *evoked culture*.[7] This is the idea that humans have cultural instincts that are triggered by the right social conditions. The propensity to develop cultures of honor when the conditions are met might be in our genes.[8] Many instincts can lie dormant until triggered. For example, parental instincts kick into gear when people have children. Similarly, some evolutionary psychologists argue that people have cultural instincts that are triggered by the right context.

It can be difficult to tell whether frequently observed features of human culture are due to evoked culture vs. convergent cultural evolution.[9] Whether honor culture originates from evoked culture or convergent cultural evolution, cultures of honor can continue due to cultural transmission long after the conditions that spurred them on have vanished.

Vertical vs. Horizontal Cultural Mismatch

As previously mentioned, cultural transmission can be vertical (i.e., passed down from one generation to another) and horizontal (i.e., spread through people within the same generation). Remnants of honor culture causing violence in the modern American south is an example of vertical cultural mismatch; outdated norms persist down through generations. Horizontal cultural mismatch can occur when norms or practices from one cultural context are introduced to other contemporary contexts for which they are ill suited. For example, the United States' campaign to bring democracy to Iraq cost vast sums of money and human life. Nevertheless, democracy failed to take root there.[10] One contributing factor could be that democracy requires the right cultural context to flourish.[11] Cultural systems that work in one context might not work in others. Memes that prove innocuous (or even useful) in one cultural setting may be harmful in others. Memes encouraging people to quit their boring jobs to pursue their dreams might be fine if people have a safety net to fall back on if things don't work out. Such memes could be troublesome if they caught on in communities without such safety nets. In general, memes that are fashionable among the rich can be harmful if they are adopted by people with fewer resources.[12]

Some memes are harmful wherever they spread. For example, suicide can spread by memetic transmission causing "copycat suicides." [13] Other memes, like handwashing, seem to be undeniably positive wherever they spread. Still other memes may be helpful in some circumstances, but harmful in others (i.e., cultural mismatch). We live in an increasingly interconnected world in which memes can be easily transmitted from one cultural context to another. More interconnectedness increases the likelihood of horizontal cultural mismatch.

Different Memes Help Different People

The previous sections focused on memes that are helpful in one cultural context, but not others. On an individual level, some memes might help one person, but harm others. Dating advice from attractive people might not work for less attractive people. The exercise routines of professional athletes might cause injuries if attempted by novices. Parenting strategies that might work well for raising one child may not work well for others (not every child has the same needs). People seem to intuit this in that they tend to copy the habits of people similar to themselves (especially the successful ones).[14] Choosing helpful memes means finding ones that right for oneself.

Conclusion

Norms and practices that evolved in one cultural context may be harmful in others. This can occur when outdated norms are passed down through generations (i.e., vertical mismatch), or when norms are transmitted from one contemporary context to another for which they are ill suited (i.e., horizontal mismatch).

Chapter Thirteen
<u>Gene-Culture Coevolution</u>

The evolution of new cultural practices can change human genes. The most-studied example of this gene-culture coevolution involves lactose tolerance.[1] Lactose is the sugar in milk, and human babies possess an enzyme called lactase that enables them to digest the lactose in their mothers' milk. After babies are weaned, they generally lack the enzyme to digest milk. Today, about 65% of the world's population becomes lactose intolerant in adulthood.[2]

In the majority of contexts throughout human history, people did not have access to milk except when breastfeeding. In such circumstances, there would be no advantages of continuing to produce lactase into adulthood. Therefore, humans evolved to produce lactase only when needed for breastfeeding. This all changed with the domestication of milk-producing animals such as cows and sheep. In cultures that began to herd such animals, the ability to gain extra calories by drinking animal milk was a huge advantage. Those individuals who had genetic mutations that made them produce lactase into adulthood were able to take advantage of this new source of nutrition. This would help them to survive and reproduce (passing on the lactase-persistence mutations to their children). On an evolutionary timescale, human lactase persistence is a relatively new phenomenon (emerging around eight thousand years ago).[3] And it resulted from culture.

Not all cultures herd milk-producing animals. The practice is most common in Northern Europe and certain regions of East Africa and the Arabian Peninsula.[4] Historically, only those cultures benefitted from lactase persistence, so only people in those cultures evolved lactose tolerance. The consequences are still felt today in that decedents from these cultures tend to have the genes for lactase persistence, but others don't. The cultural practice of herding milk-producing animals shaped human genes.

The use of fire to cook food is another cultural innovation that altered human genes. In this case, the genetic effects extend to all humans. Before our ancient ancestors could control fire (around 1.8 million years ago), they needed strong jaws and teeth to chew through raw meat and vegetation. Cooked food is easier to eat and more nutritious. When eating cooked food became common, the selection pressure for powerful jaws and teeth diminished. This led humans to develop weaker jaws and daintier teeth.[5]

Humans evolved to participate in cultural systems (i.e., the propensity for culture is in our genes). The cultural practices we develop can, in turn, shape our genes.

Chapter Fourteen
Cultural Progress

Chesterton's Fence

The writer and philosopher G.K. Chesterton said that you should never tear down a fence if you don't know the reason it was put up.[1] Practices and norms that emerge through the BVSR of cultural evolution may be helpful, even when the benefits are difficult to observe. Altering them can have unforeseeable consequences.

The psychologist Joe Henrich detailed one such example involving food preparation.[2] Manioc is a starchy tuber that is a staple crop for the Tukanoans, who are indigenous to the Colombian Amazon. It contains toxic levels of cyanide, which can cause neurological problems, paralysis, and immune suppression. However, the Tukanoans have a traditional technique of preparing manioc that removes the cyanide. The process takes days and involves washing, scraping, boiling, and letting the manioc sit. Henrich noted that it would be difficult for one individual to figure out the detoxification process. Instead, this process seems to be the result of a long process of cultural evolution.

In the 17th century, the Portuguese took manioc from South American and introduced it to parts of West Africa. The cultivation of manioc spread across Africa because it is hardy and produces high yields. However, when the Portuguese brought manioc to Africa, they ignored the traditional preparation method used by the Tukanoans. It is easy to see why. The process is arduous and the effects of chronic cyanide poisoning can take a while to become apparent. The introduction of manioc without the traditional preparation techniques led to chronic cyanide poisoning in Africa. The Portuguese would have benefitted from Chesterton's advice. Although the utility of traditional manioc preparation may not have been readily apparent, it was nonetheless crucially important.

Some cultural practices are unambiguously harmful. These are especially easy to identify with the benefit of hindsight. Ritual child sacrifice, the burning witches at the stake, and slavery are stains on human culture. Outdated norms and practices can be passed down long after they cease to be helpful (i.e., vertical cultural mismatch). Cultural evolution requires the abandonment of harmful and outdated norms and practices. However, it is often difficult to tell which cultural elements can be safely eliminated, especially when they may serve underlying functions that we don't understand.

Donald Campbell pointed out that when biologists find a perplexing form of animal life, they tend to assume that evolutionary logic lies behind any bizarre physical or behavioral characteristics. He argued that we should take a similar position with respect to cultural evolution.[3] Traditions are the result of cultural evolution, and they might contain underlying wisdom that eludes both observers and the very people transmitting them.

Overimitation

Children seem to naturally heed the wisdom of Chesterton's Fence. When they imitate adults, they tend to copy every action, even ones that seem trivial or unnecessary. This is known as overimitation.[4] For example, suppose I were to show a toddler how to put her toys away in a box. If I were to tap the toys on each side of the box before putting them away, she would likely copy that unnecessary step. When children learn procedures from adults, they may not be able to tell the difference between necessary and unnecessary steps, so the safe bet is just to copy every step. In the toy box example, the child may eventually learn that tapping the toys on the side of the box is unnecessary. However, the default tendency is to copy the process exactly as observed. According to the sociologist and physician Nicholas Christakis, overimitation is "an evolutionary adaptation that plays a fundamental role in the development and transmission of culture in our species." [5]

Another way to phrase Chesterton's Fence is that cultural practices and norms should not be abandoned unless one knows the reasons they emerged in the first place. Of course, some culturally transmitted practices may be useless or harmful. Suppose one of the traditional steps of manioc preparation did nothing to reduce the risk of cyanide poisoning, maybe even slightly increase the risk. That step might still tag along through cultural transmission because the process as a whole works. Tinkering with the process could yield further refinements. That is how cultural evolution works. Nevertheless, the idea of Chesterton's Fence is that one should assume that the products of cultural evolution emerged for a reason. Tinkering should be done with humility.

The process of eliminating aspect of culture is blind. The elimination of some norms and practices would yield a net benefit, but we don't know which ones until we try it out and observe the consequences. This should be done incrementally. We don't want to throw the baby out with the bathwater.

Second-Order Effects

Chesterton's Fence encourages caution about abandoning elements of culture. Caution is also warranted when adding new cultural elements. For any complex system, whether biological or cultural, predicting the downstream impact of introducing novel elements can be difficult. The story of the cane toad is illustrative. In the 1930s, the cane toad was brought from its native habitat in South America to Australia to control pests on sugar cane farms.[6] The cane toad later became classified as an invasive species due to its devastating impact on the local ecosystem.[7] The desired immediate consequence (sometimes known as the first-order effect) was to get rid of pests. However, those who introduced the toads to Australia failed to anticipate the downstream consequences (i.e., second-order effects). The introduction of the cane toad had a cascading effect that disrupted the local food chain, with wide-ranging effects on the ecosystem.

The second-order effects of cultural innovations can also be difficult to discern. For example, the first-order effect of dating apps is that people are able to easily connect with potential romantic partners. Researchers are still debating the second-order effects of dating apps on relationship formation and sexual behavior, and not all of them are positive.[8]

According to Chesterton's Fence, the removal of cultural norms and practices can have negative second-order effects. The first-order effect of eliminating traditional manioc preparation was that people were able to consume it without the laborious process; the second-order effect was chronic cyanide poisoning. The addition of new cultural norms and practices can also have negative second-order effects.

Overriding Cumulative Culture

In Chapter 8, I extolled the virtues of cumulative cultural innovation and the ratchet effect. That discussion might leave one with the impression that cultural changes are always (or mostly) improvements. The discussion of Chesterton's Fence, second-order effects, and the like was meant to act as a counterbalance. Cultural progress proceeds via BVSR; that requires tinkering with cultural systems. As noted, that tinkering should be done with humility.

Humans can mentally simulate the consequences of potential cultural innovations. That capacity can lead to hubris. Many ideas that seem great in one's mind prove unsuccessful when executed in the real world. People often can't anticipate the consequences of altering complex cultural systems. Cumulative cultural evolution results from incremental improvements discovered through trial and error. BVSR in mental simulations can help guide BVSR in the real world. However, people may be tempted to impose large-scale cultural changes based on ideas that haven't proven themselves in the real world.

As one example, the ideas of the Russian biologist Trofim Lysenko led to the starvation of millions of people.

Lysenko was heavily influenced by communist doctrine, which emphasized the near infinite malleability of organisms (including humans) based on environmental factors. For example, he believed that crops could be trained to grow in colder temperatures by chilling the seeds.[9] Lysenko wielded tremendous power within the Soviet government, and his ideas were widely implemented. Local farmers, who had accumulated cultural knowledge of how to grow crops, often regarded his techniques as preposterous. However, dissent from farmers and biologists was brutally punished.[10] The implementation of Lysenko's agricultural practices led to famines, which caused millions to starve.

Lysenko's ideas were the result of ideology, mental simulation, and shoddy experimentation. In contrast, the practices of local farmers were based on knowledge acquired through the trial and error in the real world. The Lysenko debacle illustrates the dangers of overriding practices that arose from cumulative cultural evolution based on unproven ideas.

Nature Constrains Culture

Nature places limits on what is possible. As discussed in the previous section, the nature of plant genetics places limits on what agricultural practices are possible. Similarly, the nature of human genetics places limits on culture. Cultural systems must align with human nature to function.

Many aspects of culture seem arbitrary. For example, human languages have different words for objects. Calling the ball of fire in the sky "the sun" works just as well as calling it "el sol." But even human language isn't entirely arbitrary. For example, the word for mother in almost all languages starts with an "m" or "n" sound and often includes simple consonant-vowel repetition, like "mama." [11] This is because those are the sounds that are easiest for infants to produce. Human biology (the infant vocal tract) affects the words cultures select to refer to moms; some accommodate human anatomy better than others. Humans are also born with psychological adaptations (i.e., human nature) which culture must accommodate.

There have been many attempts to concoct cultural arrangements that disregard human nature. Nicholas Christakis reviewed many such examples and concluded that "the great majority of these experiments have been utter flops." [12] Some attempts at cultural engineering try to break the parent-child attachment bond to allow for collective child rearing. The Israeli kibbutzim and the American communes of the 19th and 20th centuries are well-studied examples. These arrangements tend to fail because it is human nature for parents to want to take care of their own children.[13] Parents may be right to assume that their children are better off in their hands than those of non-relatives. As just one striking figure, children are *100 times* more likely to die at the hands of a stepparent than a biological parent.[14]

Humans can imagine a wide range of cultural arrangements that may seem superior to those that have been traditionally offered. However, they are destined for failure if they don't accommodate human nature. When discussing the failure of communism, the biologist E.O. Wilson quipped "Good ideology. Wrong species." [15]

In *Anna* Karenina, Leo Tolstoy wrote "happy families are all alike; every unhappy family is unhappy in its own way." This principle applies beyond families. Regarding structure of our physical bodies, there are many more ways of being dead than being alive (your heart literally needs to be in the right place).[16] Similarly, there are few ways to build functioning cultural systems, but there are many ways to build dysfunctional ones. Human nature constrains the potential range of functional cultural systems.

Culture Constrains Nature

Humans have many prosocial instincts, but nature also endowed humans with violent and selfish impulses.[17] As Aleksandr Solzhenitsyn said in *The Gulag Archipelago*, the line separating good and evil runs through every human heart. Successful cultural systems reign in human instincts that threaten social cohesion and cooperation.

Cultural norms limit the expression of harmful human impulses. For example, laws exist to thwart theft, murder, and revenge. It is human nature to seek revenge when harmed.[18] As discussed in Chapter 12, the threat of revenge can dissuade potential evildoers. However, revenge can wreak havoc on society because it tends to spiral out of control. Research on the *magnitude gap* suggests that victims view the harm done to them as more heinous than perpetrators.[19] If someone damaged my car, I would likely view it as a big deal, but the perpetrator may shrug it off as a minor offense. Because of the magnitude gap, when victims seek revenge, they often do so in a way that seems excessive to the perpetrators.[20] Perpetrators may then feel that they have a score to settle and seek revenge of their own. This can lead to a spiral of escalating revenge. These spirals of revenge can occur between individuals, but they can also occur between groups in what are often known as blood feuds. A series of small transgressions between the Hatfield and McCoy families spiraled into one of the most bloody and famous feuds in American history.[21] Cultural systems can thwart revenge by institutionalizing the doling out of punishments. This was a key insight of *Leviathan* by Thomas Hobbes. A system of laws can reign in the human impulse to seek justice through revenge.

Cultural systems also regulate innate human selfishness. Adam Smith argued in *Wealth of Nations* that market economies channel human selfishness into prosocial behavior; people offer goods and services that benefit others to ensure their own prosperity. He famously said "it is not from the benevolence of the butcher, the brewer, or the baker, that we expect our dinner, but from their regard to their own interest." Although humans evolved to participate in cultural systems, some of our innate proclivities are ill suited to cultural living. Cultures benefit from constraining them.

Conflicting Views of Human Nature

To summarize the arguments from the last two sections, culture can regulate human nature, but only within limits. Many political disagreements stem from what the economist Thomas Sowell calls a "conflict of visions" about human malleability.[22] According to the *unconstrained vision*, humans are infinitely malleable; utopia is within our grasp if only we discover the right cultural arrangements. This vision is popular among political progressives. According to the *constrained vision*, humans are inherently limited in wisdom and virtue; culture can only manage our imperfect nature. Thus, cultural progress doesn't result from changing human nature, but from cultural systems finding better ways to manage it. This view is popular among political conservatives.

While the unconstrained vision may measure a current culture system against an imagined ideal future, the constrained vision encourages measuring a cultural system against other cultures (past or present). If taken to the extreme, the constrained version may limit progress that is within our grasp. On the other hand, an extreme form of the unconstrained vision may take drastic measures in hopes of achieving a utopia. In his book titled *Evil,* Roy Baumeister identified utopian idealism as a source of violence and cruelty.[23] People may justify atrocities if they view them as necessary to bring about a perfect society. If the ends justify the means, then the more desirable the goal, the more suffering is tolerated in its service. Utopian projects have all failed, but they have cause destruction along the way.[24]

Viewing human nature through an evolutionary lens requires acknowledging some constraints on the types of cultural systems that are possible. Due to our genetic endowment, humans are not infinitely malleable. Nevertheless, there is room to debate the extent of these constraints.

The Naturalistic Fallacy

Acknowledging human nature does not require celebrating all of its attributes. That which is natural isn't necessarily good. Violence may be just as natural as parental love, but that does not give them equal moral footing. The tendency to equate "natural" with "good" is known as the *naturalistic fallacy*.[25] A simple safeguard to this type of thinking is to note that nature is full of horrors. To watch documentaries about animal behavior is to be inundated with wanton violence, rape, parents eating their children, and the like. A behavior should not be viewed as good, or even acceptable, merely because it stems from human nature. Similarly, fitness does not imply goodness; one could be efficient at passing on genes, but still be a moral monster.

The naturalistic fallacy has historically plagued some strains of evolutionary thinking. Social Darwinists like Herbert Spencer defended much human cruelty and callousness as natural manifestations of the Darwinian struggle for existence.[26] In this view, the starvation or extermination of the weak is natural, and therefore acceptable. This is odious and unnecessary. As the scientist and animal-welfare advocate Temple Grandin said "Nature is cruel, but we don't have to be."

Tinkering with Culture

People often can't predict the consequences of changes to complex cultural systems. When surveying society's shortcomings, we may be tempted to tear down existing structures and build from scratch. However, it is easier to destroy than to build. Cultural systems evolve in a slow, cumulative fashion, and they often contain logic that eludes us.

Cultural evolution requires a balance of variation and retention.[27] Excessive variation by social reformers can undermine the retention of functional aspects of culture. On the other hand, excessive retention by staunch traditionalists can undermine the variation that drives further evolution.

Cultural progress requires tinkering with norms and practices through BVSR. This tinkering is best accomplished as small scales because inevitable errors can be confined.[28] For example, Lysenko's agricultural practices wouldn't have killed millions if they were tested on a small scale and found to be faulty. Ideas that prove useful at small scales tend to spread via cultural transmission.

Successful cultural systems reign in human impulses that could threaten social cohesion. Nevertheless, there are limits as to how much culture can override our genetic endowment; culture must accommodate human nature.

Chapter Fifteen
Chaos and Blind Variation

At this point it would help to address the question of whether BVSR is all a mirage. Maybe all outcomes were predetermined at the dawn of existence, with the universe unfolding like clockwork since the Big Bang. If this were the case, then processes may only *seem* blind because we lack the proper ability to see the deterministic forces that fully account for them.

Perhaps the most fundamental scientific endeavor is to discover the causes of events and thereby gain the ability to predict future events.[1] In a deterministic universe, all events would be predictable if we had sufficient information. Determinism is typically defined as a form of causation in which the comprehensive state of the universe at any given point in time accompanied by a complete set of the laws of nature could account for the state of the universe at any other point in time.[2] If we lived in a deterministic universe, then in *principle* all outcomes could be predicted, despite the insurmountable nature of such a task in *practice*.[3] In an indeterministic universe, some events would be unpredictable in principle. In other words, an indeterministic universe would be imbued with chance. Events could unfold in different ways depending on the whims of chance.

In the title of this book, I used "chaos" as largely synonymous with chance – that which is unpredictable in principle. (Chaos just sounds cooler than chance.) However, chaos has different meanings in different contexts. Ancient Greeks used it to refer to formless primordial matter.[4] In mathematics, chaos theory describes how tiny changes to a system can yield large differences in that system's behavior. This idea has captured the popular imagination in the form of the butterfly effect: the flutter of a butterfly's wings in Argentina could cause a tornado in Texas weeks later.[5] This book is not about chaos theory. However, there is a brief discussion of chaos theory later in this section.

As I will discuss, chance occurrences are exploited by BVSR. They drive blind variation and make the process genuinely unpredictable. There is no room for chance in a deterministic universe. The available evidence suggests that we do not live in a deterministic universe.

Quantum Uncertainty

Since the 1920s, physicists have noted that entities in the quantum realm operate in ways that are unpredictable in principle. This was the point of Heisenberg's famous uncertainty principle.[6] A quantum entity such as an electron has two important properties: its position and its velocity. If you measure its position, its velocity is unpredictable; if you measure its velocity, its position is unpredictable. Thus, it is impossible to predict primary characteristics of quantum entities.[7]

According to the physicist Paul Davies, Heisenberg's uncertainty principle means that "the system is therefore indeterministic – one might say, free to choose among a range of possibilities." [8] Those of us who find this difficult to wrap our heads around are in good company. Richard Feynman famously said "I think I can safely say that nobody understands quantum mechanics." [9] Understanding the details is not necessary to grasp the take-away message: The universe is indeterministic at its most fundamental level. In the words of Steven Hawking, quantum mechanics "introduces and unavoidable element of unpredictability or randomness into science" [10]

Some physicists have argued that there may still be "hidden variables" that would show that even the quantum realm is deterministic. However, this view has been largely relegated to the "fringe of conventional physics." [11] Building on the work of John von Neumann, John Bell developed a theorem to prove the nonexistence of hidden variables in quantum mechanics. Subsequent experimentation supports quantum indeterminacy rather than hidden variables.[12]

Quantum Uncertainty Scales Up

One may wonder whether the quantum uncertainty has any bearing on the things most of us care about. In other words, is genuine uncertainty safely confined to the quantum realm? Does it scale up to affect larger systems?

Chaos theory is useful in describing how small unpredictable events in the quantum realm can have large effects on the world. The biologist, Lee Silver, provided a dramatic demonstration of this when he explained how a random quantum event may have led to the formation of the Soviet Union.[13] Cosmic rays are made of quantum particles, which occasionally knock a single atom of a DNA molecule out of place. This is the most common source of mutations. In 1818, a single mutation occurred in the sperm cell or egg cell that later fused to become Queen Victoria of England. This mutation causes hemophilia in boys. Females don't develop the disorder, but they can transmit the mutation to their children. Victoria passed the mutation on to her daughter, Alexandria. Alexandria later married Czar Nicholas II of Russia and passed on the mutation to their son, Alexis. The mutation caused Alexis to bleed uncontrollably, and doctors could not help him. Along came a faith healer named Rasputin who convinced Alexandria that he was the boy's only hope. Rasputin, who had a legendary penchant for debauchery, gained major influence with the royal family. This caused political turmoil, which came to a head when the Bolsheviks launched a revolution, took power, and formed the Soviet Union. The ramifications of this event are still felt in global politics today. An unpredictable quantum event can cause a genetic mutation that can change the world.

Randomness vs. Chance

Although genetic mutations can be caused by unpredictable quantum events, they are not totally random. *Random*, according to a strict definition, means that all possible outcomes are equally likely.[14] However, the term is often used in a broader sense (including by authors quoted in this book). Biologists have amply demonstrated that some sections of an organism's genome are more prone to mutation than others.[15] Mutator mechanisms can regulate mutation rates, and mutation rates have been found to be responsive to environmental conditions. Because genetic mutation is especially probable in some sections of the genome and under certain conditions, genetic mutation is not random. Nevertheless, the precise nature of mutations is not dictated by mutator mechanisms, environmental conditions, or any known deterministic process; and they occur independently of their fitness effects.[16] For example, there is no evidence that heat-induced mutations would lead to phenotypic changes corresponding with heat tolerance. Thus, not all possible mutations are equally likely to occur (i.e., they are not random), but which mutations will occur is unpredictable in principle (i.e., they are subject to chance).[17]

The distinction between randomness and chance is useful in describing just how "blind" BVSR-driven problem solving is. Hans Eysenck caricatured the idea of BVSR in problem solving by pointing out that we don't draw upon our knowledge of the Peloponnesian War when trying to solve a math problem.[18] He is quite right. As mentioned in Chapter 6, blind variations are not totally random. We confine our search for solutions based on the problem at hand and our previous experience.[19] Psychological adaptations can also weigh in on what strategies we may try. Variation may be subject to chance, yet confined within a set of search parameters. These search parameters may be wide or narrow depending on the degree to which we have relevant previous experience or psychological adaptations. When obvious strategies fail to work, then we may expand our search to more remote possibilities.[20] This could approach randomness, like Thorndike's cats flailing around when trapped in a puzzlebox with no clue how to escape. Not all variation is equally blind.

Completely blind variation requires that we do not receive feedback regarding whether we are getting closer to or further from finding a solution. For example, I play a game with my kids called Topfschlagen in which I place a pot upside down in the yard with candy underneath. Then, I blindfold the kids and give them wooden spoons. They have to crawl around the yard hitting their spoon on the ground until they hit the candy-containing pot. This process is completely blind in that hitting the grass does not tell them whether they are getting closer or further away from the pot. Alternatively, if I were to shout "warmer" or "colder" as they got closer or further from the pot, this would give them feedback about where to try next. In this case, their first attempt may be completely blind, but the subsequent attempts would not.

"Blind" means that one doesn't know which variations will yield a solution until they are tried and that there is an element of chance in what variations are generated. Because chance is involved, BVSR is not deterministic.

Quantum Uncertainty Drives Blind Variation

As discussed in Chapter 5, Ashby's homeostat was programmed with a set of random numbers that it used as a basis for BVSR problem solving. Systems don't need access to a table of random numbers to engage in BVSR. Chaos in the form of quantum uncertainty permeates the universe and can be harnessed as the basis for BVSR. In biological evolution, quantum events cause mutations in genes (blind variation) that are later subjected to natural selection (selective retention). Brains are also subjected to quantum uncertainty and can cause "mutations" in our thought patterns that can be subjected to selective retention. Quantum uncertainty fuels BVSR in the mind.

The brain's computing power is based on billions of neurons communicating with each other via electrical and chemical signals. Messages are relayed from one neuron to another via chemicals called neurotransmitters. These chemical messengers open and close ion channels in the neuron that receives them. The opening and closing of these channels affects the voltage of the neuron and therefore the likelihood that that it will generate an electrical impulse and "fire." If a neuron fires, then it will release neurotransmitters, and affect the likelihood that other neurons will fire, and so on. (This is obviously an extremely simplified description of neurotransmission.) Here is the point: this process can be affected by quantum-level events (e.g., the behavior of a single ion). The neuroscientist Paul Glimcher pointed out that neurons tend to rest right near the threshold of firing. He argued that the unpredictable behavior of a single calcium ion could tip the balance and determine whether a neuron fires, and that this reflects "a fundamental indeterminacy in the nervous system." [21] Accordingly, individual neurons display unpredictable variability even in tightly controlled experiments.[22] An individual "firing decision" may seem inconsequential, but as Lee Silver argued "the indeterminate firing of a single neuron can sometimes (how often is not at all clear) stimulate an entire pathway of brain activity leading to one mental state rather than another." [23] According to chaos theory, in complex systems, tiny changes can lead to large ones. In this case, instead of the flutter of a butterfly's wings leading to a tornado, we have the unpredictable behavior of individual neurons affecting one's thought patterns.

Quantum effects produce unpredictable variability in the brain. Although this variability is often referred to as noise, Björn Brembs argued that "organisms can evolve to take advantage of quantum effects." [24] Unpredictable quantum events undergird blind variations in behavior and thoughts. This inherent chaos means that nature doesn't need to equip us with a mechanism for generating blind variations – it is built in.

Modulating the degree of variability is important. No organism could function if all its behaviors and thoughts were totally random. On the other hand, complete behavioral and cognitive rigidity would make BVSR impossible. Humans and most animals increase or decrease their levels of behavioral and cognitive variability based on the situation. Variability tends to increase in novel situations, or when one is stumped in trying to solve a problem. As Brembs put it "brains are in control of variability." [25] He said that "evolution has shaped our brains to implement 'stochasticity' in a controlled way, injecting variability 'at will.'" [26] Maye and colleagues argued that brains engage in something akin to throwing dice to generate variability, but that "they have exquisite control over when, where, and how the dice are thrown." [27]

As discussed in previous sections, mental simulation enables trial and error in the mind, which is less risky and less costly than trial and error in the real world. Bad ideas can be weeded out before executing them. Therefore, high levels of variability in one's simulations are more likely to be helpful than high levels of variability in one's behavior. If the brain can modulate its own level of variability, it would benefit from turning up the dial during mental simulation.

Quantum uncertainty in the brain has been used to make many dubious metaphysical claims.[28] In this section, I only argue that it can be harnessed as a source of blind variation, analogous to the random number table programmed into Ashby's homeostat.

Quantum uncertainty is often evoked in the debate about the existence of free will. In the next section, I will briefly touch on this issue.

Free Will

Rather than start with what it would mean for a human to have free will, it might help to ask what it would mean for any broadly defined "agent" to act freely. The physicist Hans Briegel explored this idea when discussing how machines could possess free will. He argued that free actions require "room to maneuver" and that "projective simulation" allows such room.[29] Briegel's description of projective simulation in machines closely mirrors the process of mental simulation outlined in Chapter 6. The machine stores information about previous experiences in memory. Then it combines elements of previous experience in a quasi-random fashion to create simulations. Breigel noted that quantum uncertainty could be used as the source of randomness. The machine's simulations then serve to guide its future actions. Two machines with the same initial programming could develop very different characteristics through iterated projective simulation. Briegel's concept of projective simulation is an example of BVSR. As previously mentioned, Ashby's homeostat used BVSR to solve problems with which it was presented. Briegel took the application of BVSR in machines a step further to demonstrate how it could allow a machine to reprogram itself through simulation.

For some, Briegel's machine would be capable of free action; to others, that idea is preposterous. This case highlights the difficulty in defining free will. Much of the debate about whether free will exists is clouded by difficulty in finding a consensus about what would constitute a free action. Briegel's machine does meet two common criteria for free action: 1) the agent could have done otherwise, and 2) the actions are generated by the agent itself, rather than some outside force.

In a deterministic universe, no agent (whether human or machine) could ever do otherwise. All action would be the inevitable consequence of causal factors stretching back to the Big Bang. As discussed in the previous chapter, we do not live in a deterministic universe. Quantum uncertainty imbues human thoughts and actions with an element of chance. In that sense, one could have always "done otherwise" based on the whims of chance. However, as many have noted, being at the whims of chance doesn't mean one has free will. In the words of the philosopher John Searle "the indeterminacy of the quantum level is a matter of randomness, and randomness is not the same as freedom." [30] Still, Breigel and others argue that indeterminacy may afford the wiggle room for something resembling freedom.

In Breigel's framework, simulations are truly self-generated – the machine really creates them on its own. The machine is free to create simulations that allow it to shape itself and drive its future actions. This argument rest on conceptualizing random (or quasi-random) processes as part of the machine, rather than a force imposed on it. In his words "one can not separate 'the agent' from 'the randomness'... randomness plays a constitutive role in the very definition of the agent; it is so-to-speak part of its identity." [31] This is a key insight to which I will return. For Breigel, the randomness is part of the machine.

To argue whether an agent has free will, we must first define the agent. John Hopfield, most known for his work on neural networks, noted "What do we mean by 'free will?' The idea that 'I am responsible for my actions' is somehow central, but is woefully imprecise, since it involves an undefined 'I.'" [32]

"I" is difficult to define. Even in terms of a physical body, the notion of "I" can be fuzzy. The human body contains billions of symbiotic organisms of many different species. This insight led the philosopher Daniel Dennett to ponder "Am I an organism, or a community, or both? I am both – and more." [33] One continued attack on the idea of free will involves documenting the role of unconscious processes in guiding behavior. [34] An underlying assumption is that if unconscious processes guide my behavior, "I" am not really in charge. One could make the case that unconscious processes should be include in the proverbial "I." After all, as outlined in Chapter 6, conscious thought in the form of mental simulation shapes the unconscious. In any case, to argue that "I have free will" one must determine what counts as "I." In line with Briegel, I argue that the quantum variations that support our mental simulations should be included in the concept of "I." One could say that we "use" such variation, but that would be akin to saying that we use our brains to think. We don't use our brains to think, our brains are part of us. Similarly, chaos is a part of us, not imposed on us.

According to *two-stage* theories of free will, chance events in the brain act as a springboard for free action. Alternative actions are generated based on chance, which are then selected from (BVSR). The philosopher Karl Popper described his version of a two-stage theory of free will in this way "The selection of a kind of behaviour out of a randomly offered repertoire may be an act of choice, even an act of free will ... the selection may be from some repertoire of random events, without being random in its turn." [35]

In two-stage theories of free will, we generate alternatives through blind variation, and we act as the selection mechanism. But how much control do we have in shaping who we are, in shaping our own selection mechanism? Our genes and previous experience certainly play a role. However, according to Thomas Hills' neurocognitive theory of free will, we also engage in "construction of the self" through mental simulation.[36] His account of self-creation mirrors Briegel's account of how BVSR allows a machine to reprogram itself. The process resembles a feedback loop in which a system's output is used as input for future operations. The outcomes of internally generated simulations act as input for future mental operations. Simulations don't just affect how we respond to actual events, they also affect how we act in future simulations. For example, if I simulate negotiating the purchase of a car, the outcome of that simulation can affect how I will act in future simulations of negotiations. BVSR in the form of mental simulation shapes the self, which then acts as the selection mechanism for future BVSR (e.g., in problem solving).

Ashby's homeostat showed how BVSR could be used to solve a problem with a pre-programmed selection mechanism (it was programmed to selected for settings that helped it to regain equilibrium). If we grant that randomness is part of the homeostat's identity, then the homeostat was in control of its own blind variations. However, the homeostat had no control over its selection mechanism. In other words, it could not reprogram itself. On the other hand, some two-factor theories of free will argue the BVSR in the form of mental simulation can shape one's selection mechanism – shape the self.

BVSR affords us the capacity to act otherwise and to self-generate behavior. To have free will, we must be responsible for both the blind variations and the selection process. In the previous paragraphs, I argued that we shape our own selection mechanisms (i.e., we shape ourselves) through mental simulation. We are also responsible for the blind variations that undergird the whole process. The chaos that drives them is part of us, not a force imposed from the outside.

Conclusion

Quantum uncertainty means that our universe is imbued with chance at its most fundamental level. Our brains harness unpredictable quantum events as a source of blind variation. Through mental simulation, we shape ourselves – we shape our own selection mechanisms. The thoughts and actions we generate through BVSR really come from us. In that sense, they result from our own free will.

Chapter Sixteen
Physics and the Origin of the Cosmos

The preceding sections of this book showed the Darwinian principles of BVSR at work at the levels of biology, psychology, and culture. Could BVSR manifest itself at an even more fundamental level? Physicists have argued that Darwinian principles are at play in quantum mechanics and give hints about the origin of the cosmos.

Quantum Darwinism

As outlined in the previous chapter, the universe is unpredictable at its most fundamental level – that of quantum physics. However, physicists can make accurate predictions about the behavior of larger physical objects (e.g., the orbits of planets around the sun). Thus, the unpredictable quantum realm gives rise to the predictable realm of *classical* physics. Our everyday experience of the world aligns with classical physics. We can observe the precise location of an apple, and if we drop it, we can predict the speed and trajectory of its fall. According to theoretical physicist Wojciech Zurek, a Darwinian process explains how the quantum realm gives rise to the classical realm.[1]

Entities in the quantum realm have a range of potential states, but only some manifest themselves in the classical realm.[2] Zurek argued that a process similar to natural selection determines which potential states are preserved. As in biology, replication is key. When a quantum entity such as a photon interacts with the environment, it leaves an imprint in the form of information about the state it was in during the interaction (it may be in different states in different interactions). Zurek calls these imprints "informational offspring."[3] To be selected, states must make many informational offspring. It other words, they must "deposit multiple records – copies of themselves" in the environment.[4] According to Zurek, the environment acts as a "witness." The states that are witnessed most frequently by the environment are preserved, and the other potential states fade into oblivion. In this process, quantum entities don't compete with each other. Instead, multiple states of a single quantum entity compete for which one will be preserved (i.e., which one will cross the veil into the classical realm).

Zurek's theory of *quantum Darwinism* is the result of over 25 years of research at the Los Alamos National Laboratory. Some experimental evidence supports the theory.[5] Nevertheless, it is by no means universally accepted among physicists. If the theory is true, it would suggest that Darwinian processes play a key role in shaping the physical world.

The Origin of the Cosmos

Once upon a time, a famous scientist gave a public lecture in which he explained how planets orbit the sun, how the sun orbits the center of the Milky Way galaxy, and the like. After the lecture, an old woman approached the scientist and told him that his account of the universe was rubbish. The woman explained that the world is really a flat plane resting on the back of a giant tortoise. Humoring the old woman, the scientist asked "Well, what is tortoise standing on?" She replied "You are a very clever young man, but it is turtles all the way down."[6]

Pondering the origin of the cosmos is like searching for a magic turtle at the bottom on which all of existence stands. The explanatory chain must end somewhere.[7] There are physical laws, but how did they arise? Where does order come from? Many people believe that God created the laws that govern the universe. Others seem to regard the laws of physics as eternal, requiring no further explanation.[8] The following sections will address the possibility that BVSR is sufficient to explain order. In other words, BVSR is the magic turtle at the bottom of existence.

Order from Chaos

The philosopher David Hume entertained multiple ideas of how the cosmos may have arose from processes resembling

BVSR. In one version, the churning of chaos simply stumbled onto a stable arrangement.[9] Dawkins echoed this idea when he said "Darwin's 'survival of the fittest' is really a special case of a more general law of *survival of the stable*. The universe is populated by stable things." [10] Stable arrangements of matter and energy are preserved, and unstable arrangements revert back to chaos. In other words, nature selects for stable arrangements capable of preserving themselves. Chaotic interactions of matter and energy are a form of blind variation, and those arrangements that are stable are selectively retained.

Given chaos and infinite time, order will sometimes arise by dumb luck. For example, an immortal monkey continuously pounding away at a typewriter would eventually produce the works of Shakespeare.[11] Most of the monkey's output would be nonsense, but it is statistically inevitable that, given enough time, the monkey would type "All the world's a stage." Similarly, most arrangements of matter and energy would be unstable. However, given enough time, stable arrangements will arise from chaos. When stable arrangements arise, they are likely to persist. Perhaps the current order of the universe arose by dumb luck, but was stable enough to endure the last 13.8 billion years.

This account is based on BVSR *within* our universe (blind variation of arrangements of matter and energy, and selective retention of stable arrangements). The next section will address the possibility of BVSR *among* universes.

Cosmic Evolution

Hume also entertained the idea that our universe resulted from evolution on a cosmic scale.[12] The physicist Lee Smolin has provided a detailed account of how this could work.[13] He argued that universes (plural) are subject to a form of natural selection. When a star collapses, it creates a black hole. According to Smolin, the black hole induces a big bang on the other side of the event horizon, which creates a baby universe. Thus, black holes are the means by which universes replicate. The more stars a universe has, the more black holes it will generate. A baby universe retains many of the same parameters as its parent (e.g., the masses of elementary particles). However, there are small changes in these parameters akin to mutations. By chance, some of these mutations may help the baby universe to make stars, and therefore black holes. Universes that have parameters conducive to making stars have more offspring than others. Thus, this theory is based on the replication and variation of universes. Nature selects for universes that produce the most stars, because they are the best at replicating themselves. In other words, blind variation of the parameters of universes led to selective retention of those with the most stars.

Smolin admits that his theory is speculative.[14] We can't travel to the other side of black holes to test its veracity. Nevertheless, Smolin has furnished data in support of his theory. He pointed out that a universe that contains stars is extremely unlikely to arise by chance.[15] The probability of getting a universe with stars by random variation of the fundamental parameters is about 10^{229}. That is a mind-boggling number. In addition, Smolin calculated that even tiny changes to the fundamental parameters of our universe would result in fewer stars (and therefore fewer black holes.) In other words, our universe seems optimized for the creation of stars. Smolin's theory of cosmological natural selection provides an explanation for how this could happen.

Chaos and Self-Creation

In both the accounts of the origin of the cosmos outlined above, the cosmos created itself through BVSR. The cosmos displays freedom in that it could have turned out otherwise and its trajectory self-guided (i.e., it is not determined by outside forces). This mirrors the discussion of self-creation and free will in the previous chapter. One could say that the cosmos "uses" chaos to create itself. Instead, I advocate viewing chaos as part of it – just as I advocate viewing chaos as part of us.

A Rich Universe

We live in a rich, complex universe. There would be many ways for matter and energy to organize themselves in simpler ways. Clouds of gas could find a stable arrangement and persist, but then we wouldn't have stars, planets, or life.[15] How do we account for the richness of our universe?

One possibility involves BVSR within our universe. As previously discussed, chaos and eternity will eventually lead to order, and stable forms of order tend to last. Maybe our rich universe is an improbable but fortuitous result of chaotic interactions of matter and energy on an infinite timescale. Our universe may have found other less complex forms of order in the past. Complex order may be the exception rather than the rule, but we find ourselves in a particularly complex arrangement. Of course, the arrangement would have to be complex for us to be here.

Another possibility is BVSR among universes. According to Smolin's theory, nature selects for universes with many stars. Universes optimized for the creation of stars are likely to have other forms of complexity such as planets, some of which may support life. Changes to our universe's parameters wouldn't just reduce the number of stars, they would make life less likely.[16] Thus, cosmological natural selection seems to bend toward complexity – toward richness. Our universe may be just one of many complex universes.

The previous two explanations for the rich order of our universe are based on BVSR. Either our universe happened to stumble onto and preserve a rich order, or our universe is the result of an evolutionary process that bends toward richness. In both accounts, a version of BVSR is the magic turtle at the bottom of existence.

As mentioned, some view the laws of physics themselves as the magic turtle. According to this view, the laws didn't arise from BVSR, they are eternal. If this were the case, the richness of the universe would need no explanation, it is simply dictated by the laws of physics which unfold in a deterministic fashion.

Some view God as the basis of all existence; the universe is rich because it is God's will. There are multiple views of how God executes His will. Many believers hold that the universe unfolds like clockwork according to God's preordained plan. On the other hand, proponents of *process theology* (outlined by Alfred North Whitehead) believe that God's creation is an ongoing process.[17] BVSR has no place in a preordained universe, which is deterministic and based on God's eternal will. However, BVSR fits well with some varieties of process theology. God could be viewed as a selection mechanism: God created our world from primordial chaos and continues to guide it through a process of selection. Through BVSR, God fashioned the fundamental laws of nature which dictate what future events are possible, but the future is yet to be determined.

This raises the question of what God's relationship to chaos might be. Did God emerge from chaos? Are God and chaos separate, coeternal forces? All versions of process theology are unorthodox, but these two options seem especially heretical in that they downplay God's preeminence. Another option may be that chaos is a part of God. This is similar to the idea presented in Chapter 15 that chaos is part of us and acts as a springboard for our free will and creative potential. Yet another option is that God created chaos as the raw material from which to fashion the cosmos, such that chaos is not part of God, but part of His creation. Human creativity stems from our dance with chaos; perhaps God created chaos to initiate a similar dance of creation. Despite Einstein's famous quote to the contrary, perhaps God does play dice with the universe.

According to process theology, humans play a role in creation. This account sheds light on the idea that we are made in God's image, in that we as humans can shape our world though BVSR. Even if God uniformly guides existence toward goodness and beauty, humans can manifest both good and evil through their creative powers.

We will likely never know how the cosmos originated. Nevertheless, the framework of BVSR provides intriguing possibilities for those wont to ponder the question.

Something Rather Than Nothing

Chaos and eternity are sufficient to explain order.[18] Does chaos require an explanation? This is a bit like asking "why is there something rather than nothing?" As the philosopher Robert Nozick pointed out, this question seems to assume that nothingness is default state and that any deviation from nothingness requires an explanation.[19] If nothingness is viewed as the default, then chaos seems to require an explanation. However, there is no reason to assume that nothingness is the natural, default state.

Nozick detailed multiple approaches the answering the question of why there is something rather than nothing, but noted that such answers often seem "weird." [19] The simplest answer addressed by Nozick was pointed out to him by his daughter Emily when she was twelve years old. It relies on one of philosophy's oldest maxims: nothing can come from nothing (*Ex nihilo nihil fit*).[20] Nozick summarized the argument like this "If something cannot be created out of nothing, then, since there is something, it didn't come from nothing. And there never was a time when there was only nothing." [21] Thus, the natural state is for there to be something rather than nothing. If we grant that somethingness is the natural state, we still have to explain how it came to have order. BVSR is up to the task.

We know something exists, but not just something – order. The discussion of the origin of the cosmos in the previous sections shows how order can arise from chaos. Many cultures, such as the ancient Egyptians and Mesopotamians, viewed chaos as the raw material on which order is built.[22] If we take a similar approach and regard the existence as chaos as a basic assumption, then we can explore how the cosmos is shaped by BVSR.

Conclusion

The theory of quantum Darwinism suggests that a process akin to natural selection transforms the unpredictable quantum realm into the more-or-less predicable realm of classical physics. According to this theory, nature selects for quantum states that that leave the most imprints (or informational offspring) during their interactions with the environment.

BVSR provides multiple avenues for explaining the origin of the cosmos. The churning of chaos on an infinite timescale will eventually produce a stable arrangement; perhaps our universe is the result of such a process. If, as Smolin argues, universes can replicate themselves, then they may be subject to natural selection; perhaps our universe is the product of evolution on a cosmic scale. Finally, process theology affords the idea that God acts as a selection mechanism and shaped our universe from primordial chaos. The existence of chaos is a precondition for each of these possibilities; chaos is the first step toward order. We don't need to create chaos or court it, the universe is suffused with chaos.

Either the order of the universe always existed, lurking in eternal laws of physics or God's preordained plan, or it emerged through BVSR. We can observe myriad examples of BVSR producing order in our small corner of the universe. A parsimonious explanation for the origin of the order in the universe as a whole is that it arose through a similar process.

Chapter Seventeen
<u>Facing Chaos</u>

According to the horror writer H.P. Lovecraft, "The oldest and strongest emotion of mankind is fear, and the oldest and strongest kind of fear is fear of the unknown." [1] Psychological research supports the idea that people fear the unknown.[2] The existence of chaos is unnerving because it means that our world is full of uncertainty.

People often go to great lengths to alleviate feelings of uncertainty. For example, part of the appeal of conspiracy theories is that they give people a coherent explanation for complex, incomprehensible world events.[3] If one believes that a sinister cabal is behind some distressing event, that belief might be more comforting that believing that the reasons for the event are unknowable. Of course, some bad things happen due to human machinations, but many bad things happen by chance. People are motivated to find the meaning in their suffering.[4] When random suffering occurs, the only meaning is that which we people create for themselves.

The human mind was shaped by evolution to find patterns and explain events, which enables us to make predictions. Some events occur due to chance, but people are reluctant to accept chance as an explanation. This reluctance has its merits. If we attribute our hardships to bad luck rather than our own mistakes, then we may never learn from them. If scientists presume that some aspect of nature is truly incomprehensible, then they may give up trying to understand it. Thus, it helps to assume that we can understand our world. However, chaos ensures that some things are unknowable. Although the existence of chaos may be disconcerting due to the uncertainty it brings, it is a necessary precondition for BVSR.

BVSR and Order

Donald Campbell described BVSR as the only "all-purpose" explanation for order.[5] It can explain the emergence and evolution of life. BVSR undergirds trial-and-error learning, mental simulation, human creativity, and cultural evolution. Human success stems from harnessing the power of BVSR more than any other animal. It can even explain the origin of the cosmos. When new forms of order arise, they do so through BVSR.

While reflecting on the downsides of BVSR as a scientific theory, Campbell noted "The basic insight, so useful and so thrilling when first encountered, is close to being an analytic tautology rather than a synthetic description of process: if indeed variations occur which are differentially selected and propagated, then an evolutionary process toward better fit to any set of consistent selective criteria is inevitable." [6] The logic of BVSR is simple and obvious, but it does not provide a detailed explanation of any specific phenomenon.[7] For example, the details of genetic, cultural, and cosmological evolution have important differences. Genes only propagate down through generations, but memes can spread within a single generation. In genetic evolution, death befalls unfit organisms, but in Smolin's account of cosmological evolution, there is no grim reaper stalking unfit universes (some just have more offspring than others).[8] Rather than being a scientific theory like the theory of relativity or the germ theory of disease, BVSR is a broad explanatory framework that applies to myriad disciplines.

Darwin was in the dark about many of the details of his theory of natural selection. He didn't know about DNA; he knew organisms displayed variation in their physical traits, but he didn't know how they arose from genetic variation.[9] In some cases, he was flat-out wrong about the details. Nevertheless, the general evolutionary process he described has been supported by modern science and revolutionized our understanding of biology. Biologists following in Darwin's footsteps filled in many of the details. The

Darwinian principles of blind variation and selective retention have shed light on a wide range of phenomena. Specialists have begun to fill in the details of how BVSR plays out in human and animal learning,[10] economics,[11] physics,[12] and culture.[13] Further work will help to fill in the details.

BVSR provides an overarching framework that illuminates every corner of our world. It explains how order emerges from chaos. We are participants in an unending chain of creation based on BVSR. Our bodies and minds are the products of BVSR, and we carry the torch into an uncertain future - we shape our world through BVSR.

Notes

Preface
1. Darwin, 1859 (republished 1958).
2. Dennett, 1995, p. 521.

Chapter 1
1. Dawkins, 1976.
2. Cairns-Smith, 1985.
3. Dennett, 1995; also see Stetter et al., 1993.
4. Dennett, 1995; also see Arrhenius, 1908.

Chapter 2
1. Dawkins, 1976; Dawkins, 1986; Dawkins, 2010; Kelly, 2020; Stewart-Williams, 2018.
2. Campbell, 1960.
3. Hodgson & Knudsen, 2010.

Chapter 3
1. Dawkins, 1976.
2. Burnstein, Crandall, & Kitayama, 1994.
3. Pinker, 2002.
4. Hamilton, 1964.
5. Trivers, 1971.
6. Carter & Wilkinson, 2015; Carter & Wilkinson, 2016.
7. Sugiyama, 2004.
8. Trivers, 1971.
9. Lukpat, 2022.
10. Tooby & Cosmides, 2005, p. 14.
11. Dawkins, 1976.
12. Dawkins, 1982; also see Dawkins, 1986.
13. Pinker, 2002.
14. Plomin, 2019.
15. Launer, 2020; see also Segal, 2012.
16. Bouchard et al., 1990, p. 223.
17. Plomin, 2019.
18. Turkheimer, 2000, p. 160.
19. Mitchell, 2022.

20. Cashmore, 2010.
21. Bouchard et al., 1990; Plomin, 2019.
22. Plomin, 2019, p. 57.

Chapter 4
1. Pinker, 2002.
2. DeLoache & LoBue, 2009.
3. Mineka et al., 1984; Yerkes & Yerkes, 1936.
4. Tooby & Cosmides, 2005.
5. Casement, 2010.
6. Tooby & Cosmides, 2005.
7. Flinn, 2005.
8. Li et al., 2018.
9. Galway-Witham et al., 2019; Stinger, 2016.
10. Gwynne & Rentz, 1983.
11. Garcia & Koelling, 1966.
12. Barkow et al., 1992.
13. Pinker, 2002.
14. Paley, 1828; also see Dawkins, 1986.
15. Pinker, 2002.
16. Stewart-Williams, 2018.

Chapter 5
1. Cziko, 1995.
2. Cziko, 1995; also see Simonton, 1999.
3. Brembs, 2011, p. 934; also see Maye et al., 2007, p. 5.
4. Brembs, 2011.
5. Hills, 2019.
6. Interview translated by Colin McGourty for chessbase.com. "Magnus Carlsen – 'I don't quite fit into the usual schemes'" 12/22/2011.
7. Platt, 2004.
8. Ashby, 1952.
9. Campbell, 1956, p. 107.
10. Campbell, 1960.
11. Thorndike, 1898.
12. Hills, 2019.
13. Baumeister, 2005.
14. Skinner, 1953, p. 430.

15. McDowell, 2010.
16. Edelman, 1993.
17. Calvin, 1988; also see Cziko, 1995.
18. Edelman, 1993.
19. Seth & Baars, 2005.
20. Seth & Baars, 2005.
21. Edelman, 1993.
22. Calvin, 1988.
23. Cziko, 1995, p. 68.
24. Edelman, 1993, p. 115.

Chapter 6
1. Mendes et al., 2007.
2. Hills, 2019; Hirsh et al., 2012; Simonton, 1999.
3. Campbell, 1960.
4. Killingsworth & Gilbert, 2010.
5. Baumeister & Masicampo, 2010
6. Sanna, 2000.
7. Mooneyham & Schooler, 2013.
8. Killingsworth & Gilbert, 2010.
9. Wilson et al., 2014.
10. Bacon et al., 2013.
11. Baumeister, 2005.
12. Andreasen et al., 1995, p. 1583.
13. Masicampo & Baumeister, 2011.
14. Gollwitzer, 1999.
15. Bargh, 1997; Pocket, 2004.
16. Libet, 1985.
17. Baumeister & Masicampo, 2010.
18. Ottaviani & Beck, 1987.
19. Garry et al., 1996.
20. Hesslow, 2002.
21. Taylor et al., 1998.
22. Briegel, 2012; also see Hills, 2019.
23. Baumeister & Masicampo, 2010.
24. Hobson & McCarley, 1977.
25. Briegel, 2012.
26. Revonsuo, 2000.
27. Merritt et al., 1994.

28. Roepke & Seligman, 2016.
29. Pinker, 2002.
30. Bloom, 2021.
31. Diener et al., 2006.
32. Diener et al., 2006.
33. Brickman & Campbell, 1971; also see Lykken & Tellegen, 1996.
34. Brickman & Campbell, 1971
35. Mastroianni & Ludwin-Peery, 2022.
36. Kurtz & Lyubomirsky, 2008.
37. for review, see Baumeister & Masicampo, 2010.
38. Dawkins, 1976, p. 76.
39. Baumeister & Masicampo, 2010.
40. Nagel, 1974, p. 436.
41. Sedikides et al., 2003.
42. von Hippel & Trivers, 2011.
43. Pennebaker, 1997.
44. Pennebaker, 2011.
45. Pennebaker, 1997.

Chapter 7
1. Rubin, 2023, p. 157.
2. Simonton, 1999, p. 197.
3. Simonton, 1999.
4. Poincaré, 1921, p. 387.
5. Simonton, 1999, p. 29.
6. Simonton, 2003.
7. Poincaré, 1921, p. 392.
8. cited in Root-Bernstein et al., 1993, p. 339.
9. Andreasen et al., 1995.
10. Eysenck, 1993; also see Simonton, 1999.
11. cited in Simonton, 1999, p. 58.
12. Simonton, 1999.
13. cited in Simonton, 2003, p. 479.
14. Sobel & Rothenberg, 1980.
15. Simonton, 1999.
16. Ellenbogen et al., 2007.
17. Dijksterhuis & Nordgren, 2006.
18. Hills, 2019.

19. Conway et al., 2003.
20. Eysenck, 1993.
21. Guilford, 1959.
22. Eysenck, 1993.
23. Guilford, 1950; Piers et al., 1960; also see Simonton, 1999.
24. Fredrickson, 2013.
25. Wadlinger & Isaacowitz, 2006; for review, see Fredrickson, 2013.
26. Fredrickson, 2013, p. 18.
27. for review, see Baumeister, 1990.
28. Henken, 1976, p. 36.
29. Baumeister, 1990.
30. Eysenck, 1993; also see Crespi & Badcock, 2008.
31. John Dryden quoted in Eysenck, 1993, p. 159.
32. Ludwig, 1989; Ludwig, 1992; also see Simonton, 1999.
33. Campbell, 1960, p. 391.
34. quoted in Simonton, 1999, p. 27.
35. Crespi & Badcock, 2008.
36. Rothenberg, 1990.
37. Eysenck, 1993.
38. Eysenck, 1993; Simonton, 1999.
39. Karlsson, 1970.
40. for review, see Eysenck, 1993; and Simonton, 1999.
41. Simonton, 1999.
42. Linley & Joseph, 2004; also see Jayawickreme & Blackie, 2014.
43. Sareen, 2014.
44. Taleb, 2014.
45. Taleb, 2014.
46. Kouda & Iki, 2010.
47. Gabel et al., 2022.
48. Taleb, 2014.
49. Eysenck, 1993; also see Simonton, 1999.
50. Eysenck, 1993.
51. Simonton, 1999.
52. Campbell, 1960, p. 393.
53. Eysenck, 1993; Simonton, 1999.

54. Simonton, 1999, p. 76.
55. Simonton, 1977; also see Simonton, 1999.

Chapter 8
1. Baumeister, 2005.
2. Dobzhansky, 1972, p. 529.
3. Pinker, 2002; also see Dawkins, 1982.
4. Ramos et al., 2021.
5. Baumeister, 2005; Stewart-Williams, 2018.
6. Baumeister, 2005.
7. Muthukrishna & Henrich, 2016; also see Stewart-Williams, 2018.
8. Baumeister, 2005.
9. Henrich & Gil-White, 2001.
10. Trahan et al., 2014.
11. Baumeister, 2005.
12. Dennett, 1995; also see Stewart-Williams, 2018.
13. Tennie et al., 2009.
14. Tennie et al., 2009, p. 2405.
15. Nia et al., 2015.

Chapter 9
1. Dawkins, 1976.
2. Mesoudi et al., 2006.
3. Dobzhansky, 1972.
4. Dennett, 1995; Stewart-Williams, 2018.
5. Dennett, 1995, p. 365.
6. Henrich & Gil-White, 2001.
7. Singh, 2020.
8. Singh, 2020.
9. Dennett, 1995.
10. Stewart-Williams, 2018.
11. Stack, 2002.
12. Stewart-Williams, 2018.
13. Kahneman, 2011.
14. Dennett, 1995.
15. Dennett, 1995.
16. Cziko, 1995.
17. Robertson et al., 2023; Crockett, 2017.

18. Baumeister et al., 2001, p. 323.
19. Tierney & Baumeister, 2021, p. 11.
20. Appel & Richter, 2007.
21. Hovland & Weiss, 1951.
22. Hodgson & Knudsen, 2010.

Chapter 10
1. Popper, 1979, p. 261.
2. Honeycutt & Jussim, 2020.
3. Baumeister, 2005.
4. Friedman, 1953.
5. Hodgson & Knudsen, 2010.

Chapter 11
1. Bittman, 2008.
2. Mesoudi et al., 2006.
3. Dawkins, 1986.
4. Christakis, 2019.
5. Talhelm et al., 2014.
6. Talhelm et al., 2018.
7. Talhelm et al., 2014, p. 605.

Chapter 12
1. Stewart-Williams, 2018.
2. Shackelford, 2005.
3. Nisbett & Cohen, 1996.
4. Nisbett & Cohen, 1996.
5. Campbell, 1975.
6. Gul et al., 2020.
7. Barkow et al., 1992.
8. Brown & Osterman, 2012.
9. Dennett, 1995.
10. Fukuyama & McFaul, 2007.
11. Huntington, 1996.
12. Henderson, 2019.
13. Stack, 2002.
14. Singh, 2020.

Chapter 13
1. Tishkoff et al., 2017; Gerbault et al., 2011; Feldman & Laland, 1996.
2. Itan et al., 2010.
3. Stewart-Williams, 2018.
4. Stewart-Williams, 2018.
5. Christakis, 2019.

Chapter 14
1. Chesterton, 1929.
2. Henrich, 2015.
3. Campbell, 1975.
4. Nielsen & Tomaselli, 2010.
5. Christakis, 2019, p. 414.
6. Kozlov, 2021.
7. Doody et al., 2017.
8. Castro & Barrada, 2020.
9. Kolchinsky et al., 2017.
10. Christakis, 2019.
11. Lillienfeld et al., 2014; Mesoudi et al., 2006.
12. Christakis, 2019, p. 70.
13. Baumeister, 2005.
14. Daly & Wilson, 2008.
15. quoted in Getlin, 1994.
16. Dawkins, 1986.
17. Christakis, 2019.
18. Jackson et al., 2019.
19. Baumeister et al., 1990.
20. Stillwell et al., 2008.
21. Jones, 1948.
22. Sowell, 1987.
23. Baumeister, 1997.
24. Baumeister, 1997.
25. Pinker, 2002.
26. Dennett, 1995.
27. Campbell, 1975.
28. Campbell, 1975.

Chapter 15
1. Glimcher, 2005; Maye et al., 2007.
2. van Inwagen, 2008.
3. Barrow, 2007.
4. Cohn, 1995.
5. Bishop, 2017.
6. Heisenberg, 1962.
7. Davies, 1992.
8. Davies, 1992, p. 183.
9. Feynman, 1964.
10. Hawking, 1988, p. 125.
11. Stenger, 1995, p. 102.
12. Davies, 1992; Stenger, 1995.
13. Silver, 2006.
14. Campbell, 1960; Merlin, 2010.
15. Dawkins, 1976.
16. Martincorena et al., 2012.
17. Merlin, 2010.
18. Eysenck, 1993.
19. Simonton, 1999.
20. Hills, 2019; Hirsh et al., 2012.
21. Glimcher, 2005, p. 49.
22. Maye et al., 2007.
23. Silver, 2006, p. 59.
24. Brembs, 2011, p. 931; also see Hills, 2019.
25. Brembs, 2011, p. 932.
26. Brembs, 2011, p. 933.
27. Maye et al., 2007, p. 8.
28. Stenger, 1995.
29. Briegel, 2012, p. 5.
30. Searle, 2010, p. 129.
31. Briegel, 2012, p. 5.
32. Hopfield, 1994, p. 57.
33. Dennett, 1995, p. 457.
34. Bargh, 2008, p. 146.
35. Popper, 1978, p. 348.
36. Hills, 2019, p. 1.

Chapter 16
1. Zurek, 2009.
2. Davies, 1992.
3. Zurek, 2009, p. 182.
4. Zurek, 2009, p. 182.
5. Ciampini et al., 2018.
6. adapted from Hawking, 1988.
7. Davies, 1992.
8. for a relevant discussion, see Davies, 1992.
9. Hume, 1907.
10. Dawkins, 1976, p. 15.
11. Dawkins, 1986.
12. Hume, 1907.
13. Smolin, 1997.
14. Smolin, 1992.
15. Smolin, 1997.
15. Smolin, 1997.
16. Smolin, 1997; also see Dennett, 1995.
17. Viney, 2022; also see Davies, 1992.
18. Dennett, 1995.
19. Nozick, 1981.
19. Nozick, 1981, p. 16.
20. Dennett, 1995.
21. Nozick, 1981, p. 25.
22. Cohn, 1995.

Chapter 17
1. Lovecraft, 1927, p. 1.
2. Ostafin et al., 2022; also see Hirsh et al., 2012.
3. Douglas et al., 2019.
4. Frankl, 1985.
5. Campbell, 1975, p. 1105.
6. Campbell, 1960, p. 396.
7. Hodgson & Knudsen, 2010.
8. Dennett, 1995.
9. Cziko, 1995.
10. Cziko, 1995.

11. Hodgson & Knudsen, 2010.
12. Smolin, 1997; Zurek, 2009.
13. Richerson et al., 2010.

References

Andreasen, N., O'Leary, D., Cizadlo, T., Arndt, S., Rezai, K., Watkins, L., Boles Ponto, L., & Hichwa, R. (1995). Remembering the past: Two facets of episodic memory explored with positron emission tomography. *American Journal of Psychiatry, 152,* 1576-1585.

Appel, M., & Richter, T. (2007). Persuasive effects of fictional narratives. *Media Psychology, 10,* 113-134.

Arrhenius, S. (1908). *Worlds in the making.* Harper & Row.

Ashby, W.R. (1952). *Design for a brain.* Wiley.

Bacon, A., Walsh, C., & Martin, L. (2013). Fantasy proneness and counterfactual thinking. *Personality and Individual Differences, 54,* 469-473.

Bargh, J. A. (1997). The automaticity of everyday life. In R. S. Wyer, Jr. (Ed.), *Advances in social cognition: Vol. 10.* (pp. 1–61). Erlbaum.

Bargh, J. (2008). Free will is un-natural. In J. Baer, J. Kaufman, & R. Baumeister (Eds.), *Are we free? Psychology and free will* (pp. 128-154). Oxford University Press.

Barkow, J., Cosmides, L., & Tooby, J. (Eds.). (1992). *The adapted mind: Evolutionary psychology and the generation of culture.* Oxford University Press.

Barrow, J. (2007). *New Theories of Everything.* Oxford University Press.

Baumeister, R. (1990). Suicide as escape from self. *Psychological Review, 97,* 90-113.

Baumeister, R. (1997). *Evil: Inside human cruelty and violence.* Freeman.

Baumeister, R. (2005). *The cultural animal: Human nature, meaning, and social life.* Oxford University Press.

Baumeister, R., Bratslavsky, E., Finkenauer, C., & Vohs, K. (2001). Bad is stronger than good. *Review of General Psychology, 5,* 323-370.

Baumeister, R., & Masicampo, E. (2010). Conscious thought is for facilitating social and cultural interactions: How mental simulations serve the animal-culture interface. *Psychological Review, 117,* 945-971.

Baumeister, R., Stillwell, A., & Wotman, S. (1990). Victim and perpetrator accounts of interpersonal conflict: autobiographical narratives about anger. *Journal of personality and social psychology, 59*, 994-1005.

Bishop, R. (2017). Chaos. In E. Zalta (Ed.). *The Stanford Encyclopedia of Philosophy.* plato.stanford.edu/archives/spr2017/entries/chaos

Bittman, M. (2008, July 11). Why Europeans drank beer and Asians drank tea. *The New York Times.*

Bloom, P. (2021). *The sweet spot: The pleasures of suffering and the search for meaning.* HarperCollins.

Bouchard, T., Lykken, D., McGue, M., Segal, N., Tellegen, A. (1990). Sources of human psychological differences: The Minnesota study of twins reared apart. *Science, 250,* 223-228.

Brembs, B. (2011). Towards a scientific concept of free will as a biological trait: Spontaneous actions and decision-making in invertebrates. *Proceedings of the Royal Society B, 278,* 930-939.

Brickman, P., & Campbell, D. (1971). Hedonic relativism and planning the good society. In M. Appley (Ed.), *Adaptation level theory: A symposium* (PP. 287-305). New York: Academic Press.

Briegel, H. (2012). On creative machines and the physical origins of freedom. *Scientific Reports, 2,* 1-6.

Brown, R. & Osterman, L. (2012). Culture of honor, violence, and homicide. In T. Shackelford & V. Weekes-Shackelford (Eds.), *The Oxford handbook of evolutionary perspectives on violence, homicide, and war* (pp. 218–232). Oxford University Press.

Burnstein, E., Crandall, C., & Kitayama, S. (1994). Some neo-Darwinian decision rules for altruism: weighing cue for inclusive fitness as a function of the biological importance of the decision. *Journal of Personality and Social Psychology, 67,* 773-789.

Cairns-Smith, G. (1985). *Seven clues to the origin of life.* Cambridge University Press.

Cashmore, A. (2010). The Lucretian swerve: The biological basis of human behavior and the criminal justice system.

Proceedings of the National Academy of Sciences, 107, 4499-4504.

Campbell, D. (1956). Adaptive behavior from random response. *Behavioral Science, 1,* 105-110.

Campbell, D. (1960). Blind variation and selective retention in creative thought as in other knowledge processes. *Psychological Review, 67,* 380-400.

Campbell, D. (1975). On the conflicts between biological and social evolution and between psychology and moral tradition. *American Psychologist, 30,* 1103-1126.

Carter, G. G., & Wilkinson, G. S. (2015). Social benefits of non-kin food sharing by female vampire bats. *Proceedings of the Royal Society B: Biological Sciences, 282,* 20152524.

Carter, G. G., & Wilkinson, G. S. (2016). Common vampire bat contact calls attract past food-sharing partners. Animal Behavior, 116, 45–51.

Casement, A. (2010). Archetype. In D. Lemming, K. Madden, & S. Marlan (Eds.), *Encyclopedia of Psychology and Religion* (pp. 67-70). Springer.

Castro, A., & Barrada, J. (2020). Dating apps and their sociodemographic and psychosocial correlates: A systemic review. *International Journal of Environmental Research and Public Health, 17,* 6500-6525.

Calvin, W. (1988). A global brain theory. *Science, 240,* 1802-1803.

Chesterton, G. (1929). *The Thing.* S&W.

Christakis, N. (2019). *Blueprint: The evolutionary origins of a good society.* Little, Brown Spark.

Ciampini, M., Pinna, G., Mataloni, P., & Paternostro, M. (2018). Experimental signature of quantum Darwinism in photonic cluster states. *Physical Review, 98,* 1-6.

Cohn, N. (1995). *Cosmos, chaos, and the world to come: The ancient roots of apocalyptic faith.* Yale University Press.

Conway, A., Kane, M., & Engle, R. (2003). Working memory capacity and its relation to general intelligence. *Trends in Cognitive Sciences, 7,* 547-552.

Crespi, B., & Badcock, C. (2008). Psychosis and autism as diametrical disorders of the social brain. *Behavioral and Brain Sciences, 31,* 241-320.

121

Crockett, M. (2017). Moral outrage in the digital age. *Nature Human Behaviour, 11,* 769-771.

Cziko, G. (1995). *Without miracles: Universal selection theory and the second Darwinian revolution.* MIT Press.

Daly, M., & Wilson, M. (2008). Is the "Cinderella effect" controversial?: A case study of evolution-minded research and critiques thereof. In C. Crawford & D. Krebs (Eds.), *Foundations of Evolutionary Psychology* (pp. 383–400). Taylor & Francis Group; Lawrence Erlbaum Associates.

Darwin, C. (1958). *On the origin of species by means of natural selection.* New American Library. (Original work published 1859)

Davies, P. (1992). *The mind of God: The scientific basis for a rational world.* Simon & Schuster.

Dawkins, R. (1976). *The selfish gene.* Oxford University Press.

Dawkins, R. (1982). *The extended phenotype: The long reach of the gene.* Oxford University Press.

Dawkins, R. (1986). *The blind watchmaker.* W.W. Norton & Company.

Dawkins, R. (2010). Universal Darwinism. In M. Bedau & C. Cleland (Eds.), *The Nature of Life: Classical and Contemporary Perspectives from Philosophy and Science* (pp. 360-373). Cambridge University Press.

DeLoache, J., & LoBue, V. (2009). The narrow fellow in the grass: Human infants associate snakes and fear. *Developmental Science, 12,* 201-207.

Dennett, D. (1995). *Darwin's dangerous idea: Evolution and the meanings of life.* Simon & Schuster.

Diener, E., Lucas, R., & Scollon, C. (2006). Beyond the hedonic treadmill: Revising the adaptation theory of well-being. *American Psychologist, 61,* 305-314.

Dijksterhuis, A., & Nordgren, L. (2006). A theory of unconscious thought. *Perspectives on Psychological Science, 1,* 95-109.

Dobzhansky, T. (1972). Genetics and the diversity of behavior. *American Psychologist, 27,* 523-530.

Doody, J., Rhind, D., Green, B., Castellano, C., McHenry, C., & Clulow, S. (2017). Chronic effects of an invasive species on animal community. *Ecology, 98,* 2093-2101.

Douglas, K., Sutton, R., & Cichocka, A. (2019) Belief in conspiracy theories: Looking beyond gullibility. In J. Forgas & R. Baumeister (Eds.), *The Social Psychology of Gullibility: Conspiracy Theories, Fake News and Irrational Beliefs* (pp. 61-76). Routledge.

Edelman, G. (1993). Neural Darwinism: Selection and reentrant signaling in higher brain function. *Neuron, 10,* 115-125.

Ellenbogen, J., Hu, P., Payne, J., Titone, D., & Walker, M. (2007). Human relational memory requires time and sleep. *Proceedings of the National Academy of Sciences, 104,* 7723-7728.

Eysenck, H. (1993). Creativity and personality: Suggestions for a theory. *Psychological Inquiry, 4,* 147-178.

Feldman, M., & Laland, K. (1996). Gene-culture coevolutionary theory. *Trends in Ecology & Evolution, 11,* 453-457.

Feynman, R. (1964). *Quantum Mechanics.* The Messenger Lectures, MIT.

Flinn, M., Ward, C., & Noone, R. (2005). Hormones and the human family. In D. Buss (Ed.), *The handbook of evolutionary psychology* (pp. 522-580). John Wiley & Sons.

Frankl, V. (1985). *Man's search for meaning.* Simon and Schuster.

Fredrickson, B. (2013). Positive emotions broaden and build. *Advances in Experimental Social Psychology, 47,* 1-53.

Friedman, M. (1953). The methodology of positive economics. In *Essays in positive economics* (pp. 3-43). University of Chicago Press.

Fukuyama, F., & McFaul, M. (2007). Should democracy be promoted or demoted? *The Washington Quarterly, 31,* 23-45.

Gabel, L., Liphardt, A., Hulme, P., Heer, M., Zwart, S., Sibonga, J., Smith, S., & Boyd, S. (2022). Incomplete recovery of bone strength and trabecular microarchitecture at the distal tibia 1 year after return from long duration spaceflight. *Scientific Reports, 12,* 1-13.

Galway-Witham, J., Cole, J., & Stringer, C. (2019). Aspects of human physical and behavioural evolution during the last 1 million years. *Journal of Quaternary Science, 34,* 355-378.

Garcia, J., & Koelling, R. (1966). Relation of cue to consequence in avoidance learning. *Psychonomic Science, 4,* 123-124.

Garry, M., Manning, C., & Loftus, E. (1996). Imagination inflation: Imagining a childhood event inflates confidence that it occurred. *Psychonomic Bulletin & Review, 3,* 208-214.

Gerbault, P., Liebert, A., Itan, Y., Powell, A., Currat, M., Burger, J., Swallow, D., & Thomas, M. (2011). Evolution of lactase persistence: An example of human niche construction. *Philosophical Transactions of the Royal Society B, 366,* 863-877.

Getlin, J. (1994, October 21). Natural Wonder: At heart, Edward Wilson's an ant man, but it's his theories on human behavior that stir up trouble. *Los Angeles Times.*

Glimcher, P. (2005). Indeterminacy in brain and behavior. *Annual Review of Psychology, 56,* 25-56.

Gollwitzer, P. (1999). Implementation intentions: Strong effects of simple plans. *American Psychologist, 54,* 493-503.

Guilford, J. (1950). Creativity. *The American Psychologist, 5,* 444–454.

Guilford, J. (1959). Traits of creativity. In H. Anderson (Ed.), *Creativity and its cultivation* (pp. 142-161). Harper.

Gul, P., Cross, S., & Uskul, A. (2020). Applied implication of culture of honor theory and research for practitioners and prevention researchers. *American Psychologist, 76,* 502-515.

Gwynne, D., & Rentz, D. (1983). Beetles on the bottle: Male buprestids mistake stubbies for females (coleoptera). *Journal of the Australian Entomological Society, 22,* 79-80.

Hamilton, W. (1964). The genetical evolution of social behavior. 1. *Journal of Theoretical Biology, 7,* 1-16.

Hawking, S. (1988). *A Brief History of Time.* Bantam Press.

Heisenberg, W. (1962). *Physics and philosophy: The revolution in modern science.* HarperCollins.

Henderson, R. (2019, August 17). 'Luxury beliefs' are the latest status symbol for rich Americans. *New York Post.*

Henken, V. (1976). Banality reinvestigated: A computer-based content analysis of suicide and forced death documents. *Suicide and Life-Threatening Behavior, 6,* 36-43.

Henrich, J. (2015). *The Secret of Our Success: How Culture is Driving Human Evolution, Domesticating Our Species, and Making Us Smarter.* Princeton University Press.

Henrich, J., & Gil-White, F. (2001). The evolution of prestige: Freely conferred deference as a mechanism for enhancing the benefits of cultural transmission. *Evolution and Human Behavior, 22,* 165-196.

Hesslow, G. (2002). Conscious thought as simulation of behavior and perception. *Trends in cognitive sciences, 6,* 242–247.

Hills, T. (2019). Neurocognitive free will. *Proceedings of the Royal Society B, 286,* 1-9.

Hirsh, J., Mar, R., & Peterson, J. (2012). Psychological entropy: A framework for understanding uncertainty-related anxiety. *Psychological Review, 119,* 304-320.

Hobson, A., & McCarley, R. (1977). The brain as a dream state generator: An activation-synthesis hypothesis of the dream process. *The American Journal of Psychiatry, 134,* 1335-1348.

Hodgson, G.M., & Knudsen, T. (2010). *Darwin's conjecture.* The University of Chicago Press.

Honeycutt, N., & Jussim, L. (2020). A model of political bias in social science research. *Psychological Inquiry, 31,* 73-85.

Hopfield, J. (1994). Physics, computation, and why biology looks so different. *Journal of Theoretical Biology, 171,* 53-60.

Hovland, C., & Weiss, W. (1951). The influence of source credibility on communication effectiveness. *The Public Opinion Quarterly, 15,* 635-650.

Hume, D. (1907). *Dialogues concerning natural religion.* William Blackwood.

Huntington, S. (1996). *The Clash of Civilizations and the Returning of World Order.* Simon and Schuster.

Itan, Y., Jones, B., Ingram, C., Swallow, D., & Thomas, M. (2010). A worldwide correlation of lactase persistence phenotype and genotypes. *BMC Evolutionary Biology, 10,* 1-11.

Jackson, J., Choi, V., & Gelfand, M. (2019). Revenge: A multilevel review and synthesis. *Annual Review of Psychology, 70,* 319-345.

Jayawickreme, E., & Blackie, L. (2014). Post-traumatic growth as positive personality change: Evidence, controversies and future directions. *European Journal of Personality, 28,* 312-331.

Jones, C. (1948). *The Hatfields and the McCoys.* The University of North Carolina Press.

Kahneman, D. (2011). *Thinking, Fast and Slow.* Macmillan.

Kantrowitz, A. (2019, July 23). The man who build the retweet: "we handed a loaded weapon to 4-yar olds." *BuzzFeed News.* https://www.buzzfeednews.com/article/alexkantrowitz/how-the-retweet-ruined-the-internet

Karlsson, J. (1970). Genetic association of giftedness and creativity with schizophrenia. *Hereditas, 66,* 177-182.

Kelly, D. (2020). *The Origin of Phenomena* (2nd ed.). Woodhouse Press.

Killingsworth, M., & Gilbert, D. (2010). A wandering mind is an unhappy mind. *Science, 330,* 932.

Kolchinsky, E., Kutschera, U., Hossfeld, U., & Levit, G. (2017). Russia's new Lysenkoism. *Current Biology, 27,* 1037-1059.

Kouda, K., & Iki, M. (2010). Beneficial effects of mild stress (hormetic effects): Dietary restriction and health. *Journal of Physiological Anthropology, 29,* 127-132.

Kozlov, M. (2021). Australia's cane toads evolved to be cannibals at frightening speed. *Nature, 597,* 19-20.

Kurtz, J., & Lyubomirsky, S. (2008). Toward a durable happiness. In S. Lopez (Ed.), *Positive psychology: Exploring the best in people, Vol. 4. Pursuing human flourishing* (pp. 21–36). Praeger Publishers; Greenwood Publishing Group.

Launer, J. (2020). Identical strangers. *Postgraduate Medical Journal, 96,* 59-60.

Li, N., van Vugt, M., & Colarelli, S. (2018). The evolutionary mismatch hypothesis: Implications for psychological science. *Current Directions in Psychological Science, 27,* 38-44.

Libet, B. (1985). Unconscious cerebral initiative and the role of conscious will in voluntary action. *Behavioral and Brain Sciences, 8,* 529–566.

Linley, P., & Joseph, S. (2004). Positive change following trauma and adversity: A review. *Journal of Traumatic Stress, 17,* 11-21.

Lilienfeld, Scott O., Lynn, Steven J., Namy, Laura L., Woolf, Nancy J. (2014). *Psychology: From Inquiry to Understanding (Third Edition).* Pearson-Prentice Hall.

Lovecraft, H.P. (1927). *Supernatural Horror in Literature.* Dover.

Ludwig, A. (1989). Reflections on creativity and madness. *American Journal of Psychotherapy, 43,* 4-14.

Ludwig, A. (1992). Creative achievement and psychopathology: Comparison among professions. *American Journal of Psychotherapy, 46,* 330-356.

Lukpat, A. (2022, April 28). Fertility doctor accused of using his own sperm is ordered to pay millions. *The New York Times.*

Lykken, D., & Tellegen, A. (1996). Happiness is a stochastic phenomenon. *Psychological Science, 7,* 186-189.

Masicampo, E., & Baumeister, R. (2011). Consider it done!: Plan making can eliminate the cognitive effects of unfulfilled goals. *Journal of Personality and Social Psychology, 101,* 667-683.

Mastroianni, A., & Ludwin-Peery, E. (2022). Things could be better. OSF.IO/BK7ZT

Martincorena, I., Seshasayee, A., & Luscombe, N. (2012). Evidence of non-random mutation rates suggests an evolutionary risk management strategy. *Nature, 485,* 95-98.

Maye, A., Hsieh, C., Sugihara, & Brembs, B. (2007). Order in spontaneous behavior. *PLoS ONE, 2,* e443.

McDowell, J. (2010). Behavioral and neural Darwinism: Selectionist function and mechanisms in adaptive behavior dynamics. *Behavioural Processes, 84,* 358-365.

Mendes, N., Hanus, D., & Call, J. (2007). Raising the level: Orangutans use water as a tool. *Biology Letters, 3,* 453-455.

Merlin, F. (2010). Evolutionary chance mutation: A defense of the modern synthesis' consensus view. *Philosophy & Theory in Biology, 2,* 1-22.

Merritt, J., Stickgold, R., Pace-Schott, E., Williams, J., & Hobson, J. (1994). Emotion profiles in the dreams of men and women. *Consciousness and Cognition, 3,* 46-60.

Mesoudi, A., Whiten, A., & Laland, K. (2006). Towards a unified science of cultural evolution. *Behavioral and Brain Sciences, 29,* 329-383.

Mineka, S., Davidson, M., Cook, M., & Keir, R. (1984). Observational conditioning of snake fear in rhesus monkeys. *Journal of Abnormal Psychology, 93,* 355-372.

Mitchell, K. (2022). Developmental noise is an overlooked contributor to innate variation in psychological traits. *Behavioral and Brain Sciences, 45,* e171.

Mooneyham, B., & Schooler, J. (2013). The costs and benefits of mind-wandering: A review. *Canadian Journal of Experimental Psychology, 67,* 11-18.

Muthukrishna, M., & Henrich, J. (2016). Innovation in the collective brain. *Philosophical Transactions of the Royal Society B, 371,* 1-14.

Nagel, T. (1974). What is it like to be a bat? *Philosophical Review, 83,* 435-450.

Nia, H., Jain, A., Liu, Y., Alam, M., Barnas, R., & Makris, N. (2015). The evolution of air resonance power efficiency in the violin and its ancestors. *Proceedings of the Royal Society A, 471,* 1-26.

Nielsen, M., & Tomaselli, K. (2010). Overimitation in Kalahari Bushman children and the origins of human cultural cognition. *Psychological Science, 21,* 729-736.

Nisbett, R., & Cohen, D. (1996). *Culture of Honor.* Westview Press.

Nozick, R. (1981). *Philosophical explanations.* Harvard University Press.

Ostafin, B., Papenfuss, I., & Vervaeke, J. (2022). Fear of the unknown as a mechanism of the inverse relation between life meaning and psychological distress. *Anxiety, Stress, & Coping, 35,* 379-394.

Ottaviani, R., & Beck, A. (1987). Cognitive aspects of panic disorders. *Journal of Anxiety Disorders, 1,* 15-28.

Paley, W. (1828). *Natural Theology* (2nd edition). J. Vincent.

Pennebaker, J. (1997). Writing about emotional experiences as a therapeutic process. *Psychological Science, 8,* 162-166.

Pennebaker, J. (2011). *The secret life of pronouns: What our words say about us.* Bloomsbury Publishing.

Piers, E., Daniels, J., & Quackenbush, J. (1960). The identification of creativity in adolescents. *Journal of Educational Psychology, 51,* 346-351.

Pinker, S. (2002). *The Blank Slate.* Penguin.

Platt, M. (2004). Unpredictable primates and prefrontal cortex. *Nature Neuroscience, 7,* 319-320.

Plomin, R. (2019). *Blueprint: How DNA makes us who we are.* MIT Press.

Pocket, S. (2004). Does consciousness cause behavior? *Journal of Consciousness Studies, 11,* 23-40.

Poincaré, H. (1921). *The foundations of science: Science and hypothesis, the value of science, science and method* (G. Halstead, Trans.). Science Press.

Popper, K. (1978) Natural selection and the emergence of mind. *Dialectica 32,* 339–355.

Popper, K. (1979). *Objective Knowledge: An Evolutionary Approach* (Rev. ed.). Clarendon Press.

Ramos, E., Santoya, L., Verde, J., Walker, Z., Castelblanco-Martínez, N., Kiszka, J., & Rieucau, G. (2021). Lords of the rings: Mud ring feeding by bottlenose dolphins in a Caribbean estuary revealed from sea, air, and space. *Marine Mammal Science, 38,* 364-373.

Revonsuo, A. (2000). The reinterpretation of dreams: An evolutionary hypothesis of the function of dreaming. *Behavioral and Brain Sciences, 23,* 793–1121.

Richerson, P., Boyd, R., & Henrich, J. (2010). Gene-culture coevolution in the age of genomics. *Proceedings of the National Academy of Sciences, 107,* 8985-8992.

Robertson, C,, Pröllochs, N., Schwarzenegger, K., Pärnamets, P., Van Bavel, J., & Feuerriegel, S. (2023). Negativity drives online news consumption. *Nature Human Behavior, 7,* 812-822.

Roepke, A., & Seligman, M. (2016). Depression and prospection. *British Journal of Clinical Psychology, 55,* 23-48.

Root-Bernstein, R., Bernstein, M., & Gamier, H. (1993). Identification of scientists making long-term, high-impact contributions, with note on their methods of working. *Creativity Research Journal, 6*, 329-343.

Rothenberg, A. (1990). *Creativity and madness: New finding and old stereotypes.* Johns Hopkins University Press.

Rubin, R. (2023). *The creative act: A way of being.* Penguin.

Sanna, L. (2000). Mental simulation, affect, and personality: A conceptual framework. *Current Directions in Psychological Science, 9*, 168-173.

Sareen, J. (2014). Posttraumatic stress disorder in adults: Impact, comorbidity, risk factors, and treatment. *Canadian Journal of Psychiatry, 59*, 460-467.

Searle, J. (2010). Consciousness and the problem of free will. In R. Baumeister, A. Mele, & K. Vohs (Eds.), *Free will and consciousness: How might they work?* (pp. 121-134). New York: Oxford University Press.

Sedikides, C., Gaertner, L., & Toguchi, Y. (2003). Pancultural self-enhancement. *Journal of Personality and Social Psychology, 84*, 60-79.

Segal, N. (2012). *Born together reared apart: The landmark Minnesota twin study.* Harvard University Press.

Seth, A., & Baars, B. (2005). Neural Darwinism and consciousness. *Consciousness and Cognition, 14*, 140-168.

Shackelford, T. (2005). An evolutionary psychological perspective on cultures of honor. *Evolutionary Psychology, 3*, 381-391.

Silver, L. (2006). *Challenging Nature: The Clash of Science and Spirituality at the New Frontiers of Life.* HarperCollins.

Simonton, D. (1977). Creative productivity, age, and stress: A biographical time-series analysis of 10 classical composers. *Journal of Personality and Social Psychology, 35*, 791-804.

Simonton, D. (1999). *Origins of genius: Darwinian perspectives on creativity.* Oxford University Press.

Simonton, D. (2003). Scientific creativity as constrained stochastic behavior: The integration of product, person, and process perspectives. *Psychological Bulletin, 129*, 475-494.

Singh, M. (2020). Subjective selection and the evolution of complex culture. *PsyArxiv*

(doi:10.31234/osf.io/4t2ud)

Skinner, B.F. (1953). *Science and Human Behavior.* Free Press.

Smolin, L. (1992). Did the universe evolve?. *Classical and Quantum Gravity, 9,* 173-191.

Smolin, L. (1997). *The Life of the Cosmos.* Oxford University Press.

Sobel, R., & Rothenberg, A. (1980). Artistic creation as stimulated by superimposed versus separated visual images. *Journal of Personality and Social Psychology, 39,* 953-961.

Sowell, T. (1987). *A Conflict of Visions: Ideological Origins of Political Struggles.* William Morrow & Co.

Stack, S. (2002). Media coverage as a risk factor in suicide. *Injury Prevention, 8,* 30-32.

Stenger, V. (1995). *The Unconscious Quantum: Metaphysics in Modern Physics and Cosmology.* Prometheus Books.

Stetter, K., Huber, R., Blöchl, E., Kurr, M., Eden, R., Fielder, M., Cash, H., & Vance, I. (1993). Hyperthermophilic archaea are thriving in deep North Sea and Alaskan oil reservoirs. *Nature, 365,* 743-745.

Stewart-Williams, S. (2018). *The ape that understood the universe: How the mind and culture evolve.* Cambridge University Press.

Stillwell, A.M., Baumeister, R.F., & Del Priore, R. (2008). We're all victims here: Toward a psychology of revenge. *Basic and Applied Social Psychology, 30,* 253-263.

Stinger, C. (2016). The origin and evolution of Homo sapiens. *Philosophical Transactions of the Royal Society B, 371,* 1-12.

Sugiyama, L. S. (2004). Illness, injury, and disability among Shiwiar forager-horticulturalists: Implications of health-risk buffering for the evolution of human life history. *American Journal of Physical Anthropology, 123,* 371–389.

Taleb, N. (2014). *Antifragile: Things that gain from disorder.* Random House.

Taylor, S., Pham, L., Rivkin, I., & Armor, D. (1998). Harnessing the imagination: Mental simulation, self-regulation, and coping. *American Psychologist, 53,* 429-439.

Talhelm, T., Zhang, X., & Oishi, S. (2018). Moving chairs in Starbucks: Observational studies find rice-wheat cultural differences in daily life in China. *Science Advances, 4,* 1-9.

Talhelm, T., Zhang, X., Oishi, S., Shimin, C., Duan, D., Lan, X., & Kitayama, S. (2014). Large-scale psychological differences within China explained by rice versus wheat agriculture. *Science, 344,* 603-608.

Tennie, C., Call, J., & Tomasello, M. (2009). Ratcheting up the ratchet: On the evolution of cumulative culture. *Philosophical Transactions of the Royal Society B, 364,* 2405-2415.

Thorndike, E. (1898). Animal intelligence: An experimental study of the associative processes in animals. *The Psychological Review: Monograph Supplements, 2,* i-109

Tierney, J., & Baumeister, R. (2021). *The Power of Bad: How the Negativity Effect Rules Us and How We Can Rule It.* Penguin.

Tishkoff, S., Reed, F., Ranciaro, A., Voight, B., Babbitt,C., Silverman, J., Powell, K., Mortensen, H., Hirbo, J., Osman, M., Ibrahim, M., Omar, S., Lema, G., Nyambo, T., Ghori, J., Bumpstead, S., Pritchard, J., Wray, G., & Deloukas, P. (2017). Convergent adaptation of human lactase persistence in Africa and Europe. *Nature Genetics, 39,* 31-40.

Tooby, J., & Cosmides, L. (2005). Conceptual foundations of evolutionary psychology. In D. Buss (Ed.), *The handbook of evolutionary psychology* (pp. 5-67). John Wiley & Sons.

Trahan, L., Stuebing, K., Hiscock, M., & Fletcher, J. (2014). The Flynn effect: A meta-analysis. *Psychological Bulletin, 140,* 1332-1360.

Trivers, R. L. (1971). The evolution of reciprocal altruism. *The Quarterly Review of Biology, 46,* 35–57.

Turkheimer, E. (2000). Three laws of behavior genetics and what they mean. *Current Directions in Psychological Science, 9,* 160-164.

van Inwagen, P. (2008). How to think about the problem of free will. *The Journal of Ethics, 12,* 327-341.

Viney, D. (2022). Process theism. In E. Zalta (Ed.). *The Stanford Encyclopedia of Philosophy.* plato.stanford.edu/archives/sum2022/entries/process-theism von Hippel, W., & Trivers, R. (2011). The evolution and psychology of self-deception. *Behavioral and Brain Sciences, 34,* 1-56.

Wadlinger, H., & Isaacowitz, D. (2006). Positive mood broadens visual attention to positive stimuli. *Motivation and Emotion, 30,* 89-101.

Wilson, T., Reinhard, D., Westgate, E., Gilbert, D., Ellerbeck, N., Hahn, C., Brown, C., & Shaked, A. (2014). Just think: The challenges of the disengaged mind. *Science, 345,* 75-77.

Yerkes, R., & Yerkes, A. (1936). Nature and conditions of avoidance (fear) response in chimpanzee. *Journal of Comparative Psychology, 21,* 53-66.

Zurek, W. (2009). Quantum Darwinism. *Nature Physics, 5,* 181-188.